John W. Urquhart

Electric Light Fitting

A Handbook for Working Electrical Engineers

John W. Urquhart

Electric Light Fitting

A Handbook for Working Electrical Engineers

ISBN/EAN: 9783743408180

Manufactured in Europe, USA, Canada, Australia, Japa

Cover: Foto ©Lupo / pixelio.de

Manufactured and distributed by brebook publishing software (www.brebook.com)

John W. Urquhart

Electric Light Fitting

ELECTRIC LIGHT FITTING

A Handbook for Working Electrical Engineers

EMBODYING PRACTICAL
NOTES ON INSTALLATION MANAGEMENT

By JOHN W. URQUHART, Electrician
AUTHOR OF "ELECTRIC LIGHT," ETC.

WITH NUMEROUS ILLUSTRATIONS

LONDON
CROSBY LOCKWOOD AND SON
7, STATIONERS' HALL COURT, LUDGATE HILL
1890

[*All rights reserved*]

PREFACE.

THE literature of Electric Lighting is already extensive, but when regarded from the working electrician's point of view it seems to leave much to be desired. Perhaps no branch of science carried into practice has been so generally favoured with the attention of eminent mathematicians as electricity. It is more than probable that, owing to this fact, the study of electric lighting is generally presented to beginners with a bias unduly favouring the *mathematical* aspect of the question. It is, of course, well known that it is impossible to study electricity without the aid of the higher mathematics. But only a limited number of those engineers who desire to acquire the knowledge necessary for every-day purposes are acquainted with even the symbols of the calculus ; and as so many previous writers of works on electric lighting are profound mathematical thinkers, a great deal of what has already been written (and which is, no doubt, most pregnant with thought) is, for the present, not available to the average reader.

There can be little doubt, therefore, that there is a great want of teachers who will not attempt to soar above the mental capacities and attainments of men who have received only a general education .

The present volume is intended as an attempt in this direction. It consists mostly of the every-day notes of a working electrician, expressed in the simplest available language. It is addressed to intelligent men already engaged in the work of electric lighting, or training for it; and it more especially refers to the branches known as "fitting" or "wiring." The contents of the book will be found arranged more in accordance with the natural sequence of the work of electric lighting than in relation to the relative importance of the subjects. A general knowledge of electricity, and particularly of electric lighting, has been assumed on the reader's part. No attempt has been made to form a text-book, or to teach trained electrical engineers any part of their business: it is assumed that these gentlemen need no instruction from books. But some of the facts and methods dealt with in the following pages may, nevertheless, prove both new and useful even to experts.

The author gratefully acknowledges the encouragement that has been so fully accorded to his work on "Electric Light." He has also to thank those electricians who have kindly assisted him with information for the present volume, and the firms who have granted permission to publish diagrams of their latest productions.

June, 1890.

CONTENTS.

CHAPTER I.

CENTRAL STATION WORK.

PAGE

Separate Excitation of Dynamos—Series Winding—Shunt Winding—Compound Winding—Hand and Automatic Regulation—Brush's Regulator—Thomson-Houston's Regulator—Lead in the Adjustment of the Brushes—Constant Position of the Neutral Point—Notes on the Management of Dynamo Machines—Foundations—Erection of Dynamos—Speeding and Belting—Ratio of Belting Surface to Power—Lacing of Belting—Brushes—Treatment of Commutators—Asbestos—New Commutators, Fitting of—Connections of the Dynamo—Run for Mechanical Test—Hints to Dynamo Attendants—Time and Current Curve—Heat and Attrition—Overheated Armature—Suggestions and Hints—Personal Precautions—Attention to Automatic Governors   1

CHAPTER II.

LOCALISING DYNAMO FAULTS AND OBSERVATIONS RESPECTING ACCUMULATORS.

Tests for Leakage at Dynamo—Periodic Faults—Tests for Broken Conductor—Burnt-out Coils—Tests for Earth Leakage—Much Sparking at the Commutator—Example of a Rough Test for Leakage to Earth—Short Circuit or Fault in a Magnet Coil—Failure of Dynamo to Excite—Repairs to the Armature—Loose Binding—Splicing Wire—Wet Dynamo Dried by Steam—Hints to Accumulator Attendants—Best Position for Accumulators—Insulating the Accumulator—Starting and Charging an Accumulator—Working Hints—Faults in Accumulators—Automatic Switch for Accumulator—Switching in Dynamo at Right Instant—Reserve Cells 32

x   CONTENTS.

CHAPTER III.

SWITCH-BOARD AND TESTING WORK.

PAGE

Running Series or Shunt Dynamos in Parallel—Alternating Current Dynamos in Parallel—Periodicity of Alternators—Accumulators and Dynamos in Parallel—Cardew Voltmeters—Working Indicators for the Switch-room—Accumulator Hydrometer Instruments — Calibrating Voltmeter — Paterson's Voltmeter — Sir William Thomson's Voltmeters — Pocket Voltmeters — Spring Voltmeter—Ampèremeters for Station Work—Edison-Howell Lamp Indicator—Rheostats—Résumé—Resistance and Insulation Testing—Testing Box—Wheatstone's Bridge—Taking Conductor Resistance — Insulation Resistance Testing — Portable Wheatstone's Bridge—Tests During Wiring—Ordinary Conductivity Tests—Resistance Tests 52

CHAPTER IV.

ARC LIGHT WIRING AND FITTING.

Obsolete Single Arc System—Arc Lamp Coils—Single-regulating-coil Arc Lamp — Differential or Shunt-regulating Coil—Arc Lamps in Parallel—Focussing Arc Lamps—Distance between Carbons in Arc Lamps—Regulation when Running Arc Lamps in Series—Regulation when Running in Parallel—Arc Lamp Trimming—Arc Light Wiring and Fitting—Adjustment of Brush-system Lamp—Fresh Carbons—Arc Lamps in Alternating Current Circuits—Diameters of Ordinary Carbons—Arc Lamps in Series with Incandescent Lamps in Parallel—Arc-lighting Circuits—Running Leading Wires—Ground Leakage—Table of Sizes, &c., of Arc Light Cables—Arc Light Cables—Lightning Arrester—Pole and Wall Insulators—Insulation of the Leads—Naked Leads —Lightly-insulated Leads—Properly-insulated Leads—Heavily-insulated Leads—Planning a System of Mains and Feeders—Transformers or Converters—Working of Transformers—Impedance and Choking Coils—Singing of the Alternating Arc—Location of Transformers—Transformers in Parallel and Series—Main Safety Fuses for Transformers—Meters for Recording Electricity Supply—House-main Switchboard—Dynamo-room Switchboard 88

CONTENTS. xi

CHAPTER V.

WIRING FOR INCANDESCENT LAMPS.

PAGE

Systems of Wiring—Parallel or Multiple Arc—Fall of Potential or "Pressure" — Series-multiple Method — Three-wire System— Series System—Multiple-series System—Working off Transformers—Lamps in Parallel—Other Systems—Selection of a System — Alternating v. Continuous Currents — Parallel or Parallel Series—When a Dynamo is Used—Distributing-box System—Closed Loop Circuit—The Tree System — Size of Wire for the Circuits — Resistance of the Circuits — Table for the use of Incandescent Wiresmen — Wire Gauging— Wiring for Incandescent Lamps—Tests for Conductivity and Insulation—Nature of the Insulating Covering—Switching Arrangements—Main Switches—Double-pole Switches—Multiple-way Main Switches — Multiple Circuit Switches—Branch Line and Lamp Switches—Combined Switch and Cut-out—Plug and Removable Key Switches—Reversing Switch—Main Fuses and Cut-outs—Branch-wire Fuses—Particular Observation respecting Fuses—Caution respecting Fuses—Fuse Board—The Lamps and Fittings—Area Lighted—Light Absorbed by Glass Envelopes—Incandescent Lamps in General Use—Voltage of the Various Lamps—Nature and Description of the Lamps—Blackening of the Bulbs—Economical Efficiency of the Lamps—The Kilowatt—Fittings—Attachment for Portable Lamps—Methods of Running Wires—Rule-of-the-Road for Leading Wires—Cleat Wiring—Crossing Cleats—Case Wiring—Wiring of Buildings— The "Dangers" of Electricity—Tests during Wiring—Insulation Resistance—Estimation of the Power Required—The Volt—The Ohm—The Ampère—Relation of these Units to Mechanical Power—The Kilowatt Hour—Electromotive Force Required— Current Required—Methods of Jointing the Conductors—Materials Required—Instruments—General Suggestions—Soldering and Fluxes—Electric Lamps on Gas Fittings—Wiring of Fittings 124

CHAPTER VI.

INCANDESCENT LIGHTING OF SHIPS.

Dynamos—Driving—Compass Disturbance Aboard Ships—Error not Easily Detected—Tests for Compass Disturbance—Ship Wiring —Single Wire Work—Ship Fittings—Compass Electric Lamps —Suez Canal Projector 198

xii CONTENTS.

CHAPTER VII.

MISCELLANEOUS INFORMATION.

Rules of the Institute of Electrical Engineers—Conductors—Carrying Capacity—Accessibility—Insulation—Highest Permissible Temperature—Casings—Portable Lamps—Distance Apart—Inflammable Structures—Metallic Protection—Dangers from Apertures through Walls—Joints—Gas and Water Pipes—Overhead Conductors—Lightning Conductors—Metal Fastenings—Insulation Resistance—Switches—Bases—Cut-outs—Arc-Lamps—Transformers—Distance between + and − Terminals—Heating—Danger from Internal Contact 212

LIST OF ILLUSTRATIONS.

NO.		PAGE
1.	Separate Excitation of Dynamo	3
2.	Series Winding of Dynamo	3
3.	Shunt Winding of Dynamo	3
4.	Edison's Regulator of Dynamo	6
5.	Brush's ,, ,,	7
6.	Thomson-Houston Regulator	8
7.	Part of the above Regulator	9
8.	,, ,, ,,	10
9.	Current and Time Curve	27
10.	Device for Switching Dynamo in Parallel with Accumulator	50
11.	Cardew's Voltmeter	58
12.	Accumulator Voltmeter	59
13.	Contact Staff for Accumulator	60
14.	Hydrometer Staff for Accumulator	61
15.	Hydrometer for ,,	61
16.	,, ,, ,,	61
17.	Holden's Hydrometer	62
18.	Electro-Magnetic Voltmeter	64
19.	Pocket Voltmeter	65
20.	,, ,,	66
21.	Edison-Howell Lamp Indicator	68
22.	,, ,, ,,	69
23.	Testing Box	77
24.	Diagrams of Conductor and Insulation Resistance Tests	79

NO.		PAGE
25.	Diagrams of Conductor and Insulation Resistance Tests	79
26.	,, ,, ,, ,, ,,	79
27.	Portable Wheatstone's Bridge	83
28.	Brockie-Pell Arc Lamp	49
29.	,, ,, ,,	95
30.	Lightning Arrester, Thomson-Houston System	107
31.	Fluid Insulator	108
32.	Syphon for filling Fluid Insulator	109
33.	Fluid Insulator, double type	109
34.	Wall Conduit Tube Section	110
35.	,, ,, Elevation	110
36.	Diagram of Transformer	113
37.	,, ,, ,, in series	114
38.	Thomson-Houston Transformer	115
39.	Dynamo Room Switchboard for Accumulators	122
40.	Diagram of the Parallel System of Wiring	125
41.	,, explaining the Use of Feeders	127
42.	,, showing Fall of Pressure	128
43.	,, relating to Fall of Potential	128
44.	,, of Series-multiple Circuit	131
45.	,, of the Three-wire System	133
46.	,, of the Series Method	134
47.	,, of the Multiple Series Method	134
48.	,, of Transformers in Parallel	136
49.	,, ,, ,, ,,	136
50.	,, of the Parallel System of Wiring	137
51.	,, of Parallel Wiring	137
52.	,, of Closed Loop Parallel Circuit	142
53.	,, of the Tree System	143
54.	Trotter's Wire Gauge—back	148
55.	,, ,, —front	148
56.	Ordinary Wire Gauge	149
57.	Drake & Gorham's Ring-contact Switch	152
58.	,, ,, ,, Section	152

LIST OF ILLUSTRATIONS.

NO.		PAGE
59.	Woodhouse & Rawson's Double-break Switch	153
60.	Hedges' Double-pole Switch	154
61.	Woodhouse & Rawson's Multiple-way Switch	155
62.	Ring Contact Multiple-way Switch	155
63.	Woodhouse & Rawson's Accumulator Switch	156
64.	Branch Line or Lamp Switch	158
65.	Switch and Fuse combined	159
66.	Wall Connection	160
67.	Reversing Switch	160
68.	Woodhouse & Rawson's Main Fuses	161
69.	Hedges' Safety Plug	162
70.	Safety Fuse Plate	162
71.	Branch Fuse	163
72.	Main Fuse Plate Holder	164
73.	Fuse Board for Distributing Box	166
74.	Terminal Block for Distributing Box	166
75.	Hedges' Fuses for Switch Boards	167
76.	Scott's Fusible Plug	167
77.	Incandescent Lamp, B.C. pattern, Edison's	169
78.	„ „ „ Swan's	169
79.	Edison's Lamp with Bayonet Joint	172
80.	Hartnell's Lamp Reflector	173
81.	Trotter's Dioptric Shade	173
82.	Pendant Lamp with Reflector	174
83.	Double Wire Cleat	176
84.	„ „ Casing and Cover	177
85.	Double Wire Moulding	177
86.	Cornice Moulding	177
87.	Single Wire Casing	178
88.	Suez Canal Projector, front elevation	209
89.	„ section	210

BY THE SAME AUTHOR.

ELECTRIC LIGHT: Its Production and Use.

Embodying Plain Directions for the Treatment of Dynamo-Electric Machines, Batteries, Accumulators, and Electric Lamps. With numerous Illustrations. By JOHN W. URQUHART, Author of "Electroplating," "Electrotyping," etc. Third edition, carefully revised, with large additions. *Crown 8vo*, 396 *pp.* 7s. 6d. *cloth.*

"The book is by far the best that we have yet met with on the subject." [First edition.]—*Athenæum.*

"This is the third and enlarged edition of a work, concerning which there can be but one opinion. . . . The book may be described as a complete directory of the more important patents, and as a miniature *vade mecum* of the salient facts connected with electric-lighting."—*Electrician.*

"The whole ground of electric-lighting is more or less covered—accumulators, transformers, meters, &c., being referred to, illustrated, and explained in a very clear and concise manner."—*Telegraphic Journal.*

CROSBY LOCKWOOD & SON, 7, STATIONERS' HALL COURT, E.C.

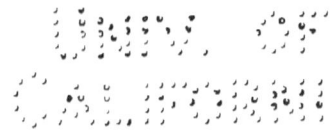

ELECTRIC LIGHT FITTING.

CHAPTER I.
CENTRAL STATION WORK.

THE duties that fall upon the electrician in charge at a central distributing station vary considerably at establishments of different capacities. But those who are training for responsible posts of this nature, whether they aspire to the care of a City central station or to the charge of a simple "installation," will find it essential to be familiar with the following leading facts and principles:—

(1.) The particular fields of application of the separately excited dynamo machine (this type of dynamo may be considered a magneto machine, as well as the obsolete permanent magnet type). The uses of the series-wound dynamo. The particular application of the shunt-wound machine. The meaning of compound winding in its two main branches of what Professor S. P. Thompson terms short-shunt compound and long-shunt compound.

(2.) How to produce *constant current* from a dynamo. How to produce constant potential. The particular application of constant current and constant potential must be known; *e.g.*, a constant potential dynamo will not necessarily run arc lamps in series, nor will a merely constant current machine run incandescent lamps in parallel.

(3.) The alternate-current dynamo, separately excited, is rising into importance, but its peculiar features are easily grasped by the student. The various methods of raising and lowering potential in this form of dynamo, by "coil grouping" and speeding, and varying the field must be familiarly known.

(4.) The nature of the magnetic circuit in a dynamo, and the meaning of "magnetic leakage" as applied to the machine.

(5.) The general management of dynamos, embodying foundation work; speeding; belting; governing, mechanically and electrically; treatment of the commutators and brushes and bearings.

(6.) Testing for faults or electrical leakage in dynamos, in mains, in branches, and in sub-branches (sometimes called "twigs").

(7.) The methods of running dynamos in parallel (particularly alternators used in incandescent lighting) and in series.

(8.) The application of voltmeters, ammeters, and other measuring instruments used in a supply station.

Separate Excitation.—Although formerly used chiefly for installations of arc lamps in series, separately excited dynamos are now largely used for incandescent lighting. In large distributing stations separately excited machines are almost exclusively (invariably so if alternators) used for feeding into the mains. The separate current is generally obtained from a smaller dynamo, series or shunt,

DYNAMO EXCITATION. 3

and sometimes compound-wound. The diagram (Fig. 1) is intended to show the disposition of the wire upon the separately excited machine. x represents the extremities of the field magnet coils, which are connected direct to the exciting machine; a shows the armature, commutator, and brushes, the current from which is led off as $+$ $-$ into the main wires of the lighting circuit.

Regulating devices of various kinds are frequently

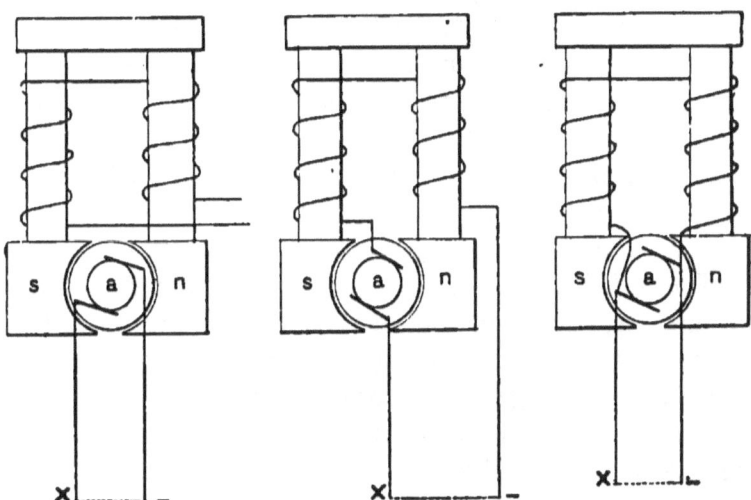

Fig. 1.—Separate Excitation.   Fig. 2.—Series Winding.   Fig. 3.—Shunt Winding

used, and placed upon the exciting machine circuit. A few of these are explained further on.

Series Winding.—For arc lamps placed in series, upon a circuit, especially if the demand for current be constant, as in street lighting, no arrangement has been found so generally serviceable as series winding. In this form of the machine the whole current from the armature circulates through the field coils. Such machines, to work satisfactorily, are generally made with comparatively light field magnets, and with numerous convolutions on the armature. Fig. 2 represents diagrammatically the course of the current in a series dynamo.

But series dynamos are very generally used upon circuits in which the number of lamps varies or the call for current is not constant. In such cases the machine is regulated by automatic devices introduced into the main circuit. One form of dynamo (Thomson-Houston) is provided with a very efficient arrangement for shifting the position of the brushes upon the commutator as the current varies, and so causing the machine to evolve more or less current as required. Another dynamo (Brush's) is provided with a regulator which shunts off the exciting part of the current from the field in such proportion as may be required.

Shunt Winding.—It has been said that the series-wound dynamo is chiefly used for arc lighting, because it is well adapted for producing constant current as distinguished from *constant potential*, which is essential in the running of incandescent lamps. The aim of a builder of dynamos for incandescent lighting is to produce a machine in which the armature resistance shall be exceedingly small. As this resistance bears a very small proportion to that of the exterior part of the circuit such a machine is found to be nearly self-regulating, especially when wound in the manner known as "*compound.*"

Fig. 3 represents the arrangement of the winding in a common shunt machine. A continuous balancing of the current goes on in such a dynamo. The current, as it is taken off the commutator by the brushes, is divided in the inverse ratio of their respective resistances between the field magnet and the exterior, or lamp, circuit. If the load of lamps increases, a larger proportion of the current passes through the shunt field coils, so strengthening the whole current. If the load of lamps be diminished, the process is the reverse of this.

Compound Winding.—For constant potential working, as in the running of incandescent lamps, a great advantage is gained by the methods of winding of the field coils known as compound. In this arrangement, which is becoming very common, two sets of coils are employed to excite the field, both series and shunt. In one form the extremities of the shunt coil are connected to the terminals of the main circuit. This is generally spoken of as a *long shunt*. In another form the shunt is connected to the brushes of the machine, known as a *short shunt*. As a rule, the series coils are wound first upon the magnet, and the shunt coils upon the outside. Sometimes the winding is the reverse of this; or they may occupy the same position with regard to the core, and lie side by side. The series coils are short and thick; the shunt coils usually long and thin. The shunt coil is usually so arranged that the machine will readily excite itself at low speed, *when the exterior portion of the circuit is open.*

Hand and Automatic Regulation.

Neither shunt nor compound winding has been found to meet the exigencies of all circuits, and the necessities of the different cases have called into use various devices for regulating the current supply to the demand. They are usually regulated by hand, or by some mechanico-electrical device upon the dynamo itself. In the management of dynamos it is essential to understand the nature of the regulator, if one be used, and we therefore select for examples three methods, now very generally employed, both in isolated plants and in central stations.

Hand Regulator.—The device shown in Fig. 4 was introduced by Edison. It consists essentially of a shunt-wound dynamo, having in the shunt portion of the circuit a rheostat R, by means of which more or less resistance can be thrown into the shunt.

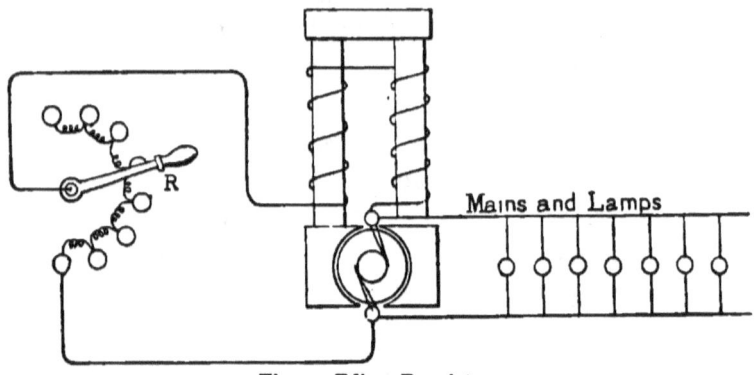

Fig. 4.—Edison Regulator.

Between each pair of studs is introduced a coil of iron wire, and, as the studs are connected in series, movement of the lever up or down will vary the length of resisting wire through which the shunted exciting current has to pass. This arrangement will be found very serviceable when the demand for current is fairly constant, or for adjusting the dynamo to a given number of incandescent lamps which are expected to be simultaneously alight. It is even in use in central distributing stations, chiefly for the incandescent lamp circuits.

Automatic Regulation.—If the demand for current is

variable, as in arc lighting or in public incandescent lighting, automatic quick-acting regulators must be used.

A very efficient form is employed by the Brush Company, chiefly for arc lighting, the simplest arrangement of which is represented as a diagram in Fig. 5. It may be used upon a series-wound dynamo. *a* represents an ordinary electro magnet (usually a pair of solenoids, with movable cores,

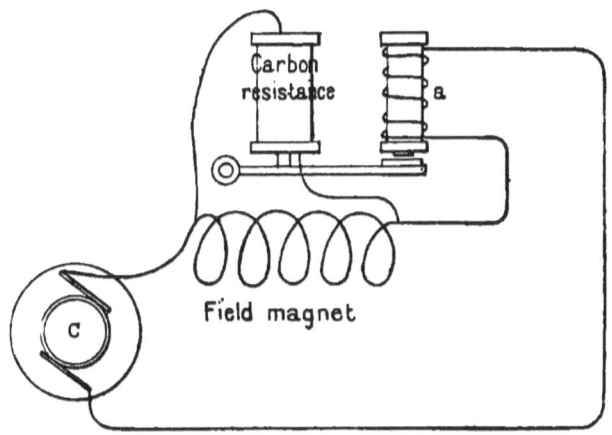

Fig. 5.—Brush's Regulator.

are used), the coil of which is in the main circuit from the dynamo *c*. When the current is normal this magnet exerts a gentle pull upon its armature. If several of the lamps in circuit become extinguished the current thereby increases rapidly. The regulator is designed to step in at this point and shunt off a portion of the current exciting the field magnets. This is effected by means of a carbon resistance column, consisting of a pile of carbon plates. In its normal position the electro magnet armature keeps the discs apart, but when the current from any cause becomes abnormally strong its pull increases, the carbon pile is compressed, and a proportion of the exciting current is thereby short-circuited through it. The effect is to weaken the field and to reduce the potential and current. The property of carbon so arranged to vary its resistance in response to a slight pressure renders this form of regulator singularly efficient.

Lately Mr. Geipel added a relay to this arrange-

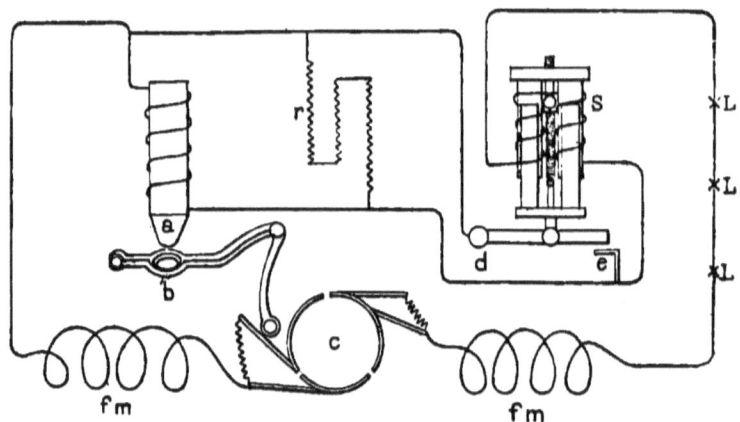

Fig. 6.—Thomson-Houston Regulator.

ment, by means of which it is rendered still more sensitive. It consists essentially of another double solenoid, controlling two contacts with the regulator solenoid wires. Upon an increase or decrease of the current taking place the relay instantly weakens or strengthens the action of the regulator coils, thus giving to the whole arrangement a double control, for while the regulator adjusts for the field magnets the relay adjusts for the regulator.

Representative of another type of automatic regulator, Fig. 6 shows as a diagram the arrangement of the

Thomson-Houston device as applied to the dynamo of that name. *a* is a straight electro magnet, with a polar extremity of conoidal form, over which the ring-like armature *bc* moves when attracted. This form of pole and armature is used also in the Thomson-Houston arc lamp. It is adapted for imparting a long pull to the armature without the liability of coming into contact.

This electro magnet, the construction of which is

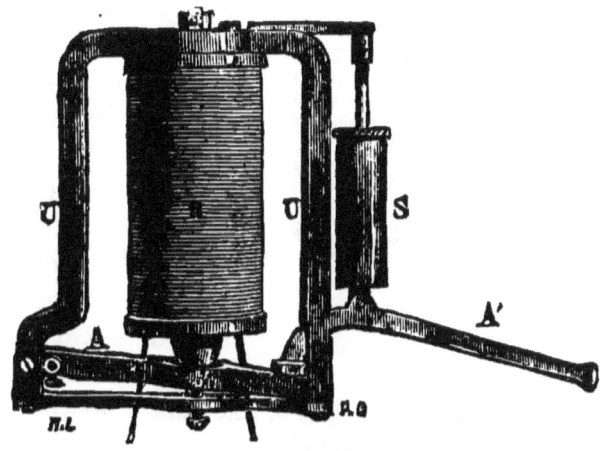

Fig. 7.—Portion of Thomson-Houston Regulator.

shown in Fig. 7, is placed in the main circuit of the machine, and its function is to *adjust the position of the brushes* upon the commutator of the machine *e*. It is well known that a change of the brushes from the normal position will result in a diminution of current. Normally the electro magnet is supposed to be short-circuited, through the by-pass wire leading to *d*, and is only brought into play by any increase or diminution of the current due to lamps being taken off or put on, or other minor causes. The electro magnet is thus controlled by the current itself by means of the electro magnetic solenoids s, which are included in any convenient part of the circuit. These solenoids are more clearly shown in Fig. 8, and their function is either to *short-circuit* the brush regulating magnet a or to put it in circuit. Referring to Fig. 8, the cores of the solenoids C are supported in position by a spring S, and they carry upon their yoke a contact point O. If the current increases in strength, the solenoids pull up the cores and break contact at O, so throwing the regulating (brush) magnet into the circuit. Reverting to Fig. 7, the lever A and A' carries a small air dash-pot S, to obviate jerky action of the parts, and in practice this lever is continuously vibrating and adjusting the brushes to the consumption of the current. The resistance r, Fig. 6, is usually composed of carbon, and is very high, its function being to absorb the destructive sparking which would otherwise occur at the contact e. The coils $f m$ represent the electro magnet of the dynamo. The Thomson-Houston commutator is distinguished by the air-blast arrangement used to blow out the sparks evolved there. This sparking cannot well otherwise be got rid of in an armature of high tension with only three parts or coils. The high tension sparking is not,

Fig. 8.—Portion of Thomson-Houston Regulator.

however, so destructive as that due to a larger current at a lower tension.

"*Lead*" *in the Adjustment of the Brushes.*—In a perfect dynamo, having no self-induction or other faults, the brushes would bear upon exactly opposite diameters of the commutator at right angles to the lines of force in the magnetic field. But in practice it is found that the most advantageous points for collection are a certain number of degrees in advance of this in the direction of motion; this is known as the *angular lead* of the brushes.

In ordinary series-wound dynamos used in arc lighting this position, once found, will generally remain constant. This is due entirely to the fact that the field magnets are in the main circuit, and any change of the current strength affects the *whole machine*, armature and field alike. But in shunt and compound-wound machines the relation does not remain constant, and the "point of best collection" may vary with the work in the circuit.

As a rule it is found that the point called neutral, or the point of *least sparking*, is the position of best collection for the brushes. In starting a dynamo this cannot be determined in any rough way until the machine is put upon full load at the normal speed. Both in relation to the output of the dynamo and to the wearing "life" of its commutator, the correct setting of the brush frame is of much importance. In well-designed machines the collecting points are always exactly opposite, and brush frames, although constructed to move a certain distance to the right and left of the collecting line, are usually fixed in relation to the diametrically opposed positions.

It will be found, notwithstanding the calculations

of several eminent electricians to the contrary, that even in dynamos with no iron in the armature, as in Siemens' alternator, or the Ferranti dynamo, "lead" must be given to the brushes. The "lead" was formerly supposed to be due to magnetic lag in the armature, but although this has undoubtedly its effect in dynamos built with iron armatures, it is not the only factor necessitating lead in the brushes.

Constant Position of the Neutral Point.—Many practical electricians suppose that the neutral point is apt to vary with the speed or output, or both. It may be of interest to note that in the experiments undertaken to determine this point by Mr. Mordey, with a Victoria dynamo and a Brush dynamo, no change in the position of the neutral point could be detected. This was true of the machines run under very different conditions of speed and load.

Notes on the Management of Dynamo Machines.

Foundations.—A great deal has been said as to the necessity for extremely solid or massive foundations. There can be no doubt that, when the dynamo itself is but flimsily constructed such a basework will be of great advantage. But when the machine is properly proportioned, and, especially, is fitted with substantial rigid brush brackets, heavy foundations are not necessary. Many dynamos perform well when merely bolted to the flooring of a factory, a sheet or two of vulcanised rubber, or better, of asbestos, being interposed. Heavy dynamos for permanent work should, of course, be carefully set upon substantial foundations.

In connection with the foundations it is interesting to note that in several small central stations, when the dynamos are placed in basements of buildings, vibration and noise are successfully combated by separating their foundations from the walls. Thus, in the Grosvenor Gallery station the foundations are heavily laid in concrete, separated from the walls of the building by a foot or so of soft clay.

The chief objection to vibration is, no doubt, the evil effect it has upon the brushes, commutator, and other collecting or regulating devices attached to the dynamo. For instance, such a dynamo as the Thomson-Houston, with its controlling apparatus, &c., would fare badly upon a light foundation.

Erecting.—Large dynamos are usually delivered from the works in parts, packed separately. In bolting the carcase together it is necessary to observe particularly that the *magnetic surfaces* (forming part of the magnetic circuit) are not only clean, but freed from oil or grease. If there is any doubt upon this point a sheet of Oakey's No. O emery cloth should be used for clearing all such surfaces. Many dynamos refuse to excite, or magnetise, on account of carelessness in erecting. Not only must such surfaces be clean, but they must *touch all over*, and no nut must be tightened up until this is ascertained. In bolting down the main castings it is necessary to avoid buckling or twisting of the frame.

Armatures are by far the most important portions of dynamos. It may be pointed out that the armature is necessarily, although heavy, a delicate part of the machine. Precautions should be taken, by means of wood packing and supports, to avoid abrasion of its wires. An accidental scratch or dent has destroyed many armatures before they were placed in the machine. If possible, *always support an armature upon its journals*, and keep it away from filings, turnings, oil, or grease. The commutator end of the armature should be protected from accidental dents or scratches. A case came under our observation in which the erecting engineer was seen to roll a heavy drum armature over an engineer's workshop floor, towards the dynamo frame. The armature had afterwards to be removed from the machine, and re-wound throughout its exterior envelope.

In erecting a dynamo it must always be borne in mind that if the running parts fit when *cold* they will become fixed when warmed up after the machine is started. End-play, to the extent of a fourth of an inch is therefore frequently allowed in armature journals, not only to allow for expansion when under load, but to assist in distributing the oil on the journals and to obviate ruts, or grooving, being started upon the surface of the commutator. For this reason, and on the score of economy of power and cool bearings, tight belting should be avoided. A belt too tight will speedily ruin a pair of journals and bearings, and will prevent end-play, with its advantages. When the armature is in position it should turn freely when moved by hand.

Speeding and Belting.—The normal speed of the dynamo is usually stamped upon it, corresponding to the volts and ampères it is estimated to yield. The driving motor should be well governed. If a gas engine is used it is a common practice to drive with a rather flexible belt, and to put a heavy balance-wheel upon the axis of the dynamo. Unsteady action of the engine or shafting will speedily be observed in pulsations, or dimming and brightening of the lamps. Gas engines that take gas once in every two or three revolutions are very troublesome on the score of "pulsating" the lights. Such engines can, however, be speeded to take gas at every revolution.

Leather belting is being displaced for the larger dynamos by rope belting in several distinct strands. This presents the advantage that a total stoppage is less likely to occur by the slipping off of the belt, or by its breaking. The ropes are run in grooved pulleys, from three to any number being used. They are undoubtedly safer, more reliable, and cheaper than one large leather belt. Most of the smaller dynamos, up to 20 h.p., are, however, fitted for belts. Riveted belts should be avoided, and joining should be done by lacing, in the old-fashioned way. The belt should always be as broad as the pulley will take, otherwise slipping, at full load, is certain to cause trouble, unless the pulley is made extra long, to allow of the belt being run off a "fast and loose" pulley gear.

Ratio of Belting Surface to Power.—The usual allowance of breadth of belt per horse-power is one inch for high-speed belting moving at the rate of 1000 feet per minute. The rule is safe for belts from three to twelve inches in width. For slower speeds a wider belt must be used.

Lacing of Belting.—Lap, switch, or splice joints are very objectionable except for large work. For high-speed driving upon small pulleys *butt* joints have proved by far the best for dynamo work. It should be noticed that a belt that emits a noisy snap upon passing over the dynamo pulley not only causes fluctuations in the light, but sets the armature, if not the whole machine, in vibration. Hence, let the belt be cut perfectly square across both ends, and laced with an endless "thong" lace. The inside face should be kept as flat as possible. New belts stretch enormously, and give a good deal of trouble in first runs. They may, therefore, be put on rather tight. Many engineers treat the harder belts with a dressing of sweet oil, frequently applied, so as to ensure pliability.

Brushes.—Each builder has his own particular pattern of brushes, and it is impossible to say which is the best form. But as to material there can be little doubt that hard-drawn or rolled copper, or phosphor bronze, gives most satisfaction in work.

Wire brushes appear to be going out of fashion. Comb-like brushes, made up from several layers of the metal, are coming generally into use. The pattern of brush sent out with a dynamo at first is generally the best for that particular machine. Two or three points may be noted—the brush should be of high conductivity; it should wear well; it should have a certain flexibility and resiliency; and it should be set in a brush bracket, *itself* provided with springs. This latter condition is of considerable importance — no commutator brush for heavy current work should be self-sprung. Only a gentle pressure upon the commutator is required; but there are two considerations that always control the amount of contact pressure. (1) In heavy, well-founded dynamos, giving currents of low tension, light pressure will be found best, because there is less vibration of the machine to cause weak contacts of the brush, and because low tension currents allow of a lighter touch without sparking: (2) For lightly-set dynamos, or those liable to vibrate, especially if giving high potential, stronger set springs are required. The snap of a badly-laced belt will frequently cause the contact to become weak periodically, producing, it may be, a burnt "spot" upon one of the commutator bars. If once such a "spot" begins it will go on from bad to worse, and finally the whole surface will need to be re-turned. Other *periodic* vibrations, perhaps due to the dynamo itself, or to adjoining machines, may start a spot. If the vibrations cannot be eliminated then more pressure must be applied at the brushes.

Let the beginner bear in mind that the pressure cannot be too light, provided efficient collection, with the minimum of sparking, occurs. The commutator and brushes are the chief care and anxiety of the electrician in charge, in the case of long runs. If he can keep them in good order, and his bearings cool, he has learnt a practical lesson of much value to him. But a burnt spot, if found persistently upon the *same commutator bar*, after re-turning, is generally due to a *fault in the armature coil connected to that bar* ; that is, the neutral line for the other coils is not the neutral line for it; the coil is out of its place in the circle, or is connected in a faulty way. We mention this in connection with brushes because it is not always bad contact at this point that originates a " bad spot."

Treatment of the Commutator.—The simplest " Commutator" is that attached to an alternating current machine, consisting as it does of a pair of copper or gun-metal rings. These are of course very easily managed. There is no liability to sparking, no burning, no production of burnt spots. The rings may be lubricated when necessary, but only lightly, and preferably with vaseline or French chalk.

A *smooth* commutator is the chief aim of the dynamo attendant. It must present neither grooves nor patches, nor parts " out of round." To attain this result, when heavy current is passing from the surface at a high speed of rotation, and for many hours together, is no easy task. But a great deal depends upon the make of the commutator itself.

Asbestos insulation between the commutator segments, which was formerly much used, gives a great deal of trouble. It easily, owing to its softness, receives into its surface copper dust or carbonised oil, and becomes a conductor, short-circuiting the bars. Various substances have been used, but experience appears to be greatly in favour of *mica;* but of this substance there are different varieties. Clean mica, free from foreign substances, and not too hard, is found to be the best for commutator insulation. When impure mica is used, or it is too hard, it does not wear away as fast as the copper, and ridges result with all their attendant trouble. The mica should wear quite as fast as the copper commutator bars. Some makers of dynamos have abandoned material insulation altogether between the bars, and have reverted to air-gaps. One instance of this is Siemens' latest dynamos, many of which have large iron commutators, insulated by air grooves. But this again will cause trouble if the grooves happen to get bridged across, an occurrence very likely with a paste of copper dust and charred oil, in long runs. In a good mica-insulated commutator there is no such trouble. We have never known mica to absorb any kind of conducting substance.

Commutators should be run without lubrication, but it is not easy to attain this. Attrition of the surface will speedily occur if there is the least roughness at first. To run a commutator dry it is necessary to have its surface even, round, and perfectly smooth— nay, burnished. *A rough surface is generally due to*

TREATMENT OF COMMUTATOR.

rough brushes. If the brush surfaces are burnished and bear upon a smooth commutator, it will be possible to run dry; but in first commencing work it is usual to slightly touch the revolving surface with oil, or, preferably, vaseline. A "touch pad" made by covering a flat piece of wood with several layers of cloth and saturated with vaseline, is very useful. This is not applied to the commutator. It is better to press the finger upon it, and transfer the layer of lubricant thus obtained to the commutator. More than a mere surface covering must be avoided. A new commutator, after a few hours' run, will under this treatment acquire a hard, brown, glossy surface, which it is very desirable to attain.

Roughness is generally treated by dressing with emery cloth. This should not be done, if it can be avoided, while the dynamo is in work. The brushes should be raised, and the No. O emery cloth wrapped around a block of wood. If these precautions are not taken, the emery powder will become embedded in the brushes, and continue to cut the surface for days thereafter. Indeed, emery, although a quick-cutting substance, should never be brought near a dynamo for this purpose. Many engineers prefer to use fine sand-paper or a leather pad with grindstone dust glued thereon.

For spots or grooves there is no effective remedy but turning in the lathe. Files are very often used, but it is quite impossible to thereby produce a true cylinder.

Large dynamos are now very frequently furnished with an accessory in the form of a miniature lathe, by means of which the commutator can be "trued" without removal from the machine. It is probable that in future all large machines will be thus wisely equipped.

A very useful device has been suggested for this purpose by Mr. R. Tatham, who proposes to furnish the brush brackets with a slow to and fro motion in line with the axis of the dynamo, and to attach to the bracket a trueing tool or emery wheel for occasional correction. But although the reciprocating motion of the brushes themselves, as proposed, would no doubt be an advantage in itself, the gear for that purpose, consisting of a worm-wheel and tangent worm shaft, would be likely to introduce faults of contact or insulation in practical use. Whatever turning device is employed for turning in position it will be found necessary to run the armature at a slow speed. In the larger stations little machines are used both for this purpose and for trimming off the ends of brushes, especially that form in which contact is made by a bundle of wires or slips.

In the case of dynamos in which regulation is effected by rocking the brushes to and from the neutral line, the commutators are apt to give much greater trouble. There is usually more sparking, which cannot be avoided. It would appear that the Thomson-Houston dynamo does not suffer much from this cause, although, owing to the high tension employed upon the arc machines and the nature of the three-coil armature, there is a good deal of sparking. But the air-blast used in this instance both serves to keep the commutator cool and clean and to extinguish the sparks.

New Commutator.—Many of the best dynamos are accompanied as an accessory with a spare commutator, which can be fitted in place of the old by

FITTING OF NEW COMMUTATOR.

observing particularly the method and order of connecting it to the armature wires. In removing the old wires, which are generally screwed to the bars, a "tally" or numbered tag should be tied to each wire, indicating exactly its position in respect to the bars of the commutator. The work of re-connecting is simple in cases where the wires are joined direct to the bars, and are not carried either to the rear or in advance of their positions upon the armature; but in many dynamos, *e.g.*, the Edison-Hopkinson type, the wires are taken 85 degrees to the rear, and there attached to the commutator. This method is adopted to allow of the *neutral* points—collecting lines—being placed in convenient positions for observation and adjustment of the brushes. Thus, instead of the points of collection being upon a vertical line, which would place the lower brush directly under the commutator, the line is nearly horizontal, and both brushes can be equally well observed.

The same method is adopted in some of the Siemens' dynamos, but in many of the best machines, where there is any likelihood of confusion in connecting, either the wire extremities are furnished with a stamped (numbered) plate for connection, or a diagram of the positions is obtainable.

Soldered connections are the most troublesome. The unsoldering is a tedious process. We may suppose the armature to be removed from the frame and placed upon supports at a convenient height. After clearing off the dust, &c., each joint should be touched with a drop of the zinc chloride solution used for soldering, and a pretty hot soldering bit applied to the spot. As soon as the solder runs the wire is lifted up, and the old solder wiped off its end. The work may be done with a mouth blowpipe. For this purpose a gas jet, attached to a rubber tube without a burner will be found a convenient source of heat. The blowpipe flame can be directed accurately upon the joint and the work done very quickly. In resoldering the commutator must first be securely keyed upon the shaft and the bars at the connecting points scraped clean. Each point should then be touched with soldering fluid (Baker's is esteemed the best) and thoroughly "tinned" with the copper bit. It must be observed by those not acquainted with the use of a copper bit that the point must be freshly filed, and, while yet bright, the solder—previously moistened with fluid—applied. The bit itself must be thoroughly tinned, and after re-heating the point should be wiped clean. In resoldering the wire ends the wire is placed, without tendency to spring, upon the tinned commutator plate. A touch of the fluid is applied (be sparing in the use of this) and a drop of solder taken up by the bit applied to the joint. It should immediately run freely and make a clean, perfect joint. It is well to run on a little more of the solder by way of a strengthener. No difficulty need be experienced if the surfaces are *clean*, the copper bit *well tinned*, and *hot* enough to cause the tin to run freely. Joints made with resin as a flux are doubtless to be generally preferred, but the use of resin is not so easily acquired, and an imperfect joint is more likely to result in inexperienced hands. There is no objection to the use of Baker's fluid if sparingly used and each joint afterwards wiped clean. It may be pointed out that the careless use of common soldering fluid is very apt to leave joints that will become rotten, or waste away by electrolysis under the influence of the current.

CONNECTIONS OF THE DYNAMO.

In the case of screwed connections to the armature plates too much attention cannot be given to the preparatory cleaning of the points of contact, and to ascertain that each screw is tight enough in its hole to ensure its holding. If the screw feels loose in its hole while screwing up, it will soon work slack, and cause an arc to form, burning the contact. For this reason some of the later machines have both screwed and soldered connections, and in some cases silver solder, applied with borax as a flux and the blowpipe, is employed.

Connections of the Dynamo. — In erecting new dynamos the connecting of the field coils to the circuit of the armature, or otherwise, is sometimes a difficult point. It will first be necessary to ascertain exactly what type of machine the dynamo is represented to belong to. If a series-wound dynamo, a separately excited, or a shunt machine, the connection can be ascertained by reference to Figs. 1, 2, 3, p. 3. But certain symbols are generally used to distinguish the extremities of the wires and the terminals. Thus, + and —, positive and negative, are widely used to indicate the "feeding" and "receiving" ends of a coil, or terminals. White (or bright) terminals are also used for +, or positive, and black terminals (representing earth) for —, or negative. The terminals are frequently spoken of as live or leading for positive, and return or earth for negative. In connecting up a dynamo two positives are never connected together, nor two negatives.

After ascertaining the particular nature of the machine, its connections, if not numbered, will depend upon the direction of rotation. If the field magnet be connected up in a series machine so that the current flows in the magnet so as to increase its residual magnetism it will be correct. Every dynamo magnet has a certain residual magnetism when the machine is at rest, and it may be desirable to ascertain which pole is N. and which S. This can readily be determined by means of a compass needle or a small magnet, for the N. pole of the dynamo will not attract the N. pole of the magnet, but it will strongly attract the S. pole, and *vice versâ*. The course of the current in that magnet will then be easily found according to the following rule :—

If a spiral of wire be taken, and a piece of iron inserted therein, and a current caused to flow in that wire in the direction of the hands of a watch, when the spiral is looked at end on, the pole of that iron nearest you is the S. pole.

Or, more simple still, If you look at a right-handed screw, the thread representing the current, the end viewed is the S. pole. This pole is, as applied to compasses and galvanometers, frequently called the "blue pole," from the custom of makers to leave the south-seeking pole blue and to brighten up the north-seeking pole.

In a separately excited machine the direction of the current in the field magnet should be particularly ascertained, otherwise the machine will yield — (negative) at its + (positive) terminal, and give rise to all kinds of trouble in the work of wiring for lamps.

If a mistake has been made, it may be rectified by strongly magnetising the field magnet by the passage of a current either from another dynamo or a battery of accumulators. This will have the effect of breaking down the residual magnetism and reversing its polarity.

When the dynamo is first started the current should be tested for direction by the use of a compass. Place a compass upon the ground; run a wire from the + pole of the dynamo over the compass and back (through a suitable resistance) to the − pole; if, while you stand with your back to the dynamo, the N. pole of the compass turns to your left hand, the current is flowing from the dynamo towards you, and is correct in respect to the positive terminal.

In a shunt, series, or compound machine, not much harm can result from starting it when wrongly connected with respect to the field magnet—it will refuse to excite, and will give no current.

When there is a resistance in the exterior portion of the circuit, and the dynamo refuses to excite, the fault is usually due to wrong connections. But a series dynamo will not excite readily at a *low* speed.

An ordinary compound (series and shunt) dynamo is connected correctly when the ends of the shunt (fine wire) coil are joined to the brushes, and the series (thick wire) coil joined, one end to the − (negative) brush, and the other to the − (negative) terminal. The current in both coils must of course flow in one direction around the magnet.

Run for Mechanical Test.—A run of several hours' duration should be made with a new dynamo to test the bearings, lubrication, stretch belts, &c. Hot bearings may gradually cool down if new after a few hours further running, but if there is any question of the armature shaft being out of alignment the heat will increase. The surface of the armature must be quite clear of the magnet, and in line with its bore.

All kinds of suggestions and substances have been recommended at different times as a cure for hot bearings, but, provided the journals be lubricated, the fault itself must be got rid of. The chief causes of heating are doubtless, (1) belting too tight; (2) bearing too short for the work; (3) badly fitted, out of round, binding, or out of alignment. The lubrication should be of heavy oil or other lubricator of good quality. In long runs with heavy dynamos the bearings sometimes become so heated as to need the application of the hose—in cases where the load of the dynamo cannot be switched on to another machine and the current must be maintained. Hot bearings are of course made hotter by the current in the wires of the armature.

In the use of needle lubricators the needles frequently stick in their tubes, owing to foreign substances in the oil. The needles should be tested for free play before starting a long run. A hot bearing, perhaps to the extent of slight "seizing" or attrition, is generally brought about by neglecting this precaution. The semi-solid lubricants, fed from suitable spring lubricators, and which flow gently when warm, are being much used for dynamos. The chief fault to guard against is failure of the lubrication while the dynamo is left by itself for long periods of time.

Notes for Dynamo Attendants.

The attendant should understand his machine. One attendant can manage several dynamos if they do not call for much regulation. In isolated stations, where the demand for electricity is constant or nearly so, compound machines will be found to regulate themselves, and the exciting current once determined and applied, need not be varied.

HINTS TO DYNAMO ATTENDANTS. 27

But in central or "public" stations the call for current varies enormously. Taking a representative incandescent lighting station as an example, the diagram (Fig. 9) shows, beginning at noon and till 3 P.M. very little demand for current. Between 3 and 4 o'clock the demand rises rapidly from 10 to 150 units; by 5 o'clock it has risen to 350 units; at 6 o'clock to 500 units; and at 7 reaches a maximum of 600 units. It then gradually drops, until, at 2 A.M., there is practically no demand.

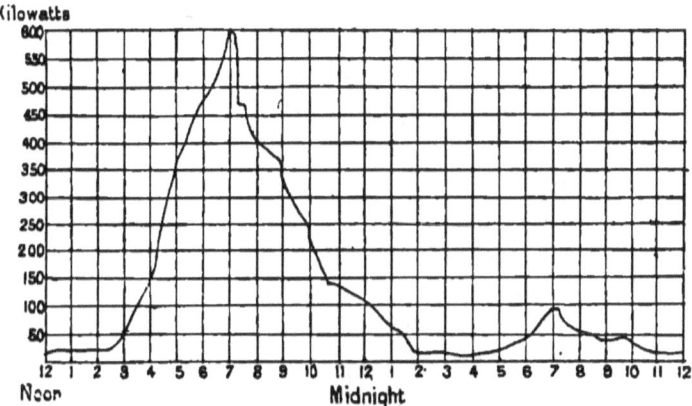

Fig. 9.—Time and Current Curve.

The requirements of such a station call for several dynamos to meet this varying demand. But one dynamo, with a good hand or automatic regulator, will be found to carry the scale up or down by itself for a considerable distance. In most central stations this regulation has hitherto been effected by means of observation and the hand.

It will be evident that the dynamo, to meet the ever varying conditions and continue the supply, must be well looked after. A few observations upon the main points likely to call for special attention are therefore offered here, in the hope that they may prove useful not to trained electricians, who are supposed to be well versed in all the best methods of meeting a varying demand, but to members of that large class who are at present serving, as it were, a kind of apprenticeship to the business of electric lighting.

Heat and Attrition.—The dynamo attendant's bugbear is doubtless *heat*. Under a high speed the bearings are apt to get hot, and under a heavy load of lamps the armature and field magnets frequently get so heated that they cannot well be touched by the hand. There is therefore under these conditions a constant danger, or supposed danger, of "firing" the bearings or journals and burning the insulation of the wire coils.

Attrition, or cutting friction of the commutator, is another cause of trouble in long runs, but is more easily overcome than the overheating.

Bearings can be kept cool if at first well-fitted, if not too short, if not binding in the "neck," and if lubricated freely with a good oil or other lubricant.

Dry cutting of the commutator is due to rough treatment, rough brush surfaces, or grit, or emery, or to *too much pressure of the springs*. The larger dynamos are fitted with several brushes upon each arm of the rocker, and a brush that is cutting should be at once taken out. The pair should be taken out even while the dynamo is running at full load. Their roughened ends should be cut off and smoothed, then, if possible, *burnished*, using a brass finisher's steel burnisher for the purpose. This will impart a glass-like surface to the copper. The commutator should then be wiped clean with a pad of wash-leather, and its rough surface smoothed, either with fine sand-paper or with emery cloth. But the use of the latter is not recommended. The smooth surfaces may then be lightly covered with the merest film of vaseline or oil, and the brushes replaced. A gentle pressure should be applied at first. When it is not possible to remove the brushes thus by instalments dry cutting should be at once stopped, as far as practicable. For this purpose, clean off the surface and apply either a flat pad covered with fine glass-paper, or, if that is not at hand, a chip of emery cloth, No. O. When as smooth as possible clean off and touch with vaseline. When the dynamo is stopped examine the pressure of the brush springs; it will generally be found that the attrition was due to this cause, in excess. The brushes should be removed and trimmed and bur-nished at the first opportunity. Avoid lubrication of commutator, if you can do so. Too much oil or vaseline will cause long, circular sparks to leap from segment to segment, greatly weakening the current.

Heated Armature and Field Coils.—*When a well-designed dynamo heats too much it is overloaded.* Lamp after lamp has been switched in until the current has become too much for the wires. It would be better to see the belt slipping than the coils becoming over-heated. But, in respect to overheating, it is not always synonymous with overloading. In many of the earlier dynamos this excess heat is due to " eddy currents" set up in the iron of the armature, owing to its imperfect subdivision. A very instructive in-stance of the enormous advantage of subdivision may be referred to. The Brush dynamo, with "solid" armatures, as first introduced, and in use in this country until very recently, will, taking one specific size, the 16-lighter, when fitted with the new laminated armature, give 25 lights in each case without overheating.

In a central station the only way to effectively reduce the heat of an overloaded dynamo is to switch in another machine, which will take half, or a proportion, of the load. The heated coils will then gradually begin to cool down. In the early days of electric lighting a fan was used to keep the dynamo cool.

It is part of the duty of the electrician in charge, more than that of the dynamo attendant, to ascertain the full working load of his machines, and to issue instructions for a watch to be kept upon the indicators, so as to determine when the current is reaching a maximum, and it is time to switch on a fresh machine. It is scarcely necessary to remind the attendant that water cannot be used to cool a heated dynamo.

Hints and Suggestions.—Keep iron and steel tools away from the machine; never file iron near to a dynamo. Lubricate with a brass, copper, or zinc oil can. Leave your watch at home, if it is an ordinary watch. Have a pair of bellows for blowing away all dust, especially metallic dust, from armature coils. Do not spill oil or water near to or upon a machine. Prevent by shields adjacent machinery from throwing oil upon the dynamo.

Tighten all binding screws afresh every day. Test nuts and bolts occasionally for tightness. If a binding screw is loose, examine it for burned surfaces, and file off a fresh surface. Adjust brushes as to pressure before starting, and as to attaining the neutral point —point of no sparking—after starting. Never lift a brush while the current is on—you would make a burned patch upon the commutator.

PERSONAL PRECAUTIONS. 31

Personal Precautions.—Never close a circuit of any dynamo, or, indeed, any circuit, through your body. It may be done by inadvertence, and not through the hands only. Many of the earlier electricians received severe shocks by merely touching a wire or terminal with one hand. This is generally due to *ground leakage*, and a pair of rubber overshoes would prevent it. But, as a general rule, never touch a wire or terminal, either with one or both hands, or with any metallic article, while current is on. If terminals need attention, an insulated key or spanner must be used, or put on a pair of thick rubber gloves.

All spanners, plyers, and adjusting tools in general, used in a dynamo room, or switch room, should have insulated handles of ebonite, or other good insulator.

Shocks of enormous tension—probably over 1000 volts—have been freely taken by many boasting persons, but it may be pointed out that the deadly nature of the electricity is not its tension merely, but the quantity or current passing, combined with high tension. A discharge of many thousand volts can easily be taken upon the knuckles of the hand, without injury, from a Leyden jar, but the same tension in a cable carrying a current of 1000 ampères would not only burn the hand but be pretty certain to kill the adventurous experimenter.

Attention to Automatic Governors.—The general impression among electricians has hitherto been that an automatic governor for public incandescent lighting is not reliable, and that hand governors only are to be depended upon, but this impression is wearing away. It is necessary, however, for the attendant to keep an eye upon both current and governor, no matter of what design, during the whole period of the run.

CHAPTER II.

LOCALISING DYNAMO FAULTS, AND OBSERVATIONS RESPECTING ACCUMULATORS.

A DYNAMO may sometimes unaccountably refuse to excite and to start. If separately excited it may refuse to give any current. This is the greatest of all faults, but it may be due to a serious defect, or simply to a very small fault, easily remedied.

But by far the most frequent complaint is due to *partial* failure of current; to fluctuations, usually sudden, in the current strength, and to occasional unaccountable extinctions of the lights. Pumping, or pulsating of the lights is another fault sometimes met with. In arc lighting extinctions and rapid self-re-lighting sometimes occur.

Broadly speaking, faults, save strictly mechanical defects, easily traced by the engineer, are usually due to *defective insulation* or *defective conduction* in the dynamo or its accessories.

A coil in the armature may be burnt, *e.g.*, the insulation charred; the commutator may not have all its sections insulated; there may be conduction, or leakage, between some part of the armature circuit with the iron body of the machine, or the field magnet coils may be similarly leaking. Possibly a small arc has been established at some point by a failure of the insulation, and this may become active or inactive, according to the current or E. M. F. of the machine.

The *periodic faults* are by far the most troublesome to detect, especially those that do not occur at half load, but appear at or near full load. Others again occur at a given speed, and are not to be traced when the armature is moving at any other speed.

The only effective means of localising faults is a system—a comprehensive system—of tests, which will broadly include all faults that have hitherto been observed. Many faults that appear to be electrical are really due to bad engineering.

Tests for Leakage to Ironwork of Machine.—The first and most important condition of the insulation is its completeness with respect to the iron body of the armature or field magnets. For practical purposes the body of the dynamo, *e.g.*, the field magnet framing and base, may be considered one conductor, complete as to conductivity. The ironwork of the armature is also generally in one with the ironwork of the frame, but it is a great advantage to have the iron core work of the armature insulated from the shaft, and therefore completely isolated. Hence, if a leakage occurs from an armature coil, it cannot get further than the core. In making a test of short circuit to the ironwork, therefore, it is not always correct to assume that the armature core is one with it.

The whole dynamo is usually insulated from the earth. If the ironwork be in contact with earth, any leakage from either armature or field coil will cause an *earth fault*.

In dynamo work ironwork fault and earth fault are usually synonymous, and may be considered together.

The testing instrument is usually a simple galvanometer. The source of current is often the dynamo itself, this test being taken while it is running. If the dynamo be standing, a few accumulator cells are the most suitable. But it is practicable to make fair tests with any of the portable or "dry" batteries now so common. For heavy dynamo work a potential of about ten volts is, however, very generally used. The instruments—galvanometer, &c.—are generally kept at a suitable distance from the dynamo, especially if it be running.

Connect one screw of the galvanometer to earth by a wire to a gas or water pipe, or other convenient "ground"; lead another wire to the dynamo, and, if it be running, giving current, any leakage of that current to the ironwork, and from the latter to earth, can be ascertained by contact of the wire with the frame of the machine. A deflection of the galvanometer would thus show a *double fault*—leakage of coil to iron, thence to earth. If the dynamo be standing idle, connect the testing battery in circuit with the galvanometer. If no deflection is obtained, and whether the dynamo is running or idle, connect battery and galvanometer to earth as before, and make contact with the wire to the ironwork of the machine. A deflection of the galvanometer will indicate that the ironwork is leaking to earth.

If the dynamo is standing, touch the commutator with the wire—a deflection of the needle will indicate that the wire coils are leaking to ironwork. If the machine be a separately excited dynamo, test the terminals of the field magnet also.

If the ironwork be found insulated, a leakage from coils to ironwork can be detected by connecting the galvanometer and battery to the frame, as to earth, and testing by coil contact as before.

A test should be taken by connecting to iron core

of armature, if it be easily accessible, as to earth, and making contact to the commutator segments. Contact to each of those should be made in succession.

It may be pointed out that if the dynamo be running while making tests, it may only be practicable to ascertain earth insulation, and a false conclusion may be drawn from them owing to a fault to earth in some part of the circuit of the lamps, removed from the machine. A fault of this latter kind would cause all lamps *beyond* the leak to burn dimly.

Tests for Internal Broken Conductors.—The continuity of the field magnet circuit is easily ascertained. Make a circuit of galvanometer, battery, and field terminals—*no deflection* will indicate a rupture of the wire, at some point—machine idle. Localising broken armature wire is also comparatively simple. When the coil extremities can be traced, or are known, make a circuit between the commutator bars attached to those extremities (remove brushes meanwhile); no deflection will indicate a break. This break is very frequently just at, or near the point of junction with the commutator.

When the armature winding is not known, and it is impossible to determine the extremities of the coils, the test for a break is not so simple. As a rule the ends of the coils are in connection with diametrically opposite segments. In this case it may only be necessary to make a circuit by touching these segments with the two wires, from galvanometer and battery. If a *weak* deflection is obtained, it may be due to one of two causes, either the insulation material between the segments has become conductive, by impressed copper dust or charred oil—a liability very common in cases of asbestos insulation—or there is a partial break or broken wire in partial contact within the coil.

When the fault cannot be located by either of these methods the armature wires should be disconnected entirely from the commutator and each other. In doing this numbers should be attached to ends and segments, indicating the connections in re-attaching. *The extremities of a coil* can then be found by touching with the test wires. If there is a pair from which no deflection can be obtained, the assumption is that the fault is in that coil.

A diagram of the armature winding as applied to the particular make of dynamo used should be kept by the attendant, and referred to when any question of a fault arises. This will indicate where to apply the test wires, and may save many disconnections and experiments.

Intermittent contacts between broken junctions are very troublesome. They will generally give a deflection upon being tested, and cannot easily be located, unless they occur at the point of contact with an armature segment. In the case of a small armature an intermittent contact was found in one case by testing each coil, and while the needle of the galvanometer remained deflected setting the armature in vibration by striking the end of the shaft with a *copper* hammer (to obviate mechanical injury). When the faulty coil came under the test the needle oscillated, showing intermittent contact between (as was found) two ends of a wire bent to a sharp angle near to the end of the armature coil.

Burnt-out Coils.—When an armature coil makes, by fault, a short circuit within itself, *e.g.*, does not deliver its current to the lamps, its current will become

TEST FOR EARTH LEAKAGE. 37

abnormally strong; it will become heated, and finally the insulation will be burnt off. This occurrence is generally amply indicated, unless it be very gradual, by *smoke arising from the armature* and a *smell of burning varnish and cotton.*

But unless in central stations, where the dynamo is constantly watched, a coil generally burns out without being observed, and the attendant is apprised of it by a dimming of the light or by fluctuations.

Short circuits are, however, when a dynamo is looked after, generally detected before burning out occurs. They are not easy to locate, especially in armatures of low resistance. The best way to determine whether a short circuit exists is to *measure the resistance* of each coil in succession; the faulty coil will then upset the balance of the test by its lower resistance. Measurement or balancing with the Wheatstone bridge, in order to detect faults in electric lighting stations or circuits, receives some little attention in the succeeding chapter, where also will be found some account of the instruments used for ordinary tests.

All the *ordinary* faults that occur in dynamos can be detected and localised by means of comparatively rough and ready methods, some examples of which have been given.

Much sparking at the commutator is generally a sign of overloading, or a short circuit in the armature.

When a dynamo shows much sparking, and begins to heat rapidly, there is usually a short circuit in the leads feeding the lamps, and this should be seen to at once, otherwise the building may be set on fire.

Example of a Rough Test for Leakage to Earth.—

In central station work, where large currents are evolved, the following rough test for detecting leakage or ground fault is very common. Two lamps are connected in series across the terminals of the dynamo. If it be a potential of 100 volts, two 100-volt lamps are used. The connecting wire between the lamps is put to earth. If there is any leakage it will be shown by the lamp connected to the terminal upon whose line the leakage exists becoming brighter each time the earth contact is made. This not only serves to indicate the lead from which the leak is to be found, but roughly, by the brightness of the lamps, its extent. Such a leakage is called a ground fault, or shortly, a ground. This is a very convenient test, not only in respect to earth leakage, but for leakage to adjacent metallic bodies.

Short Circuit or Fault in a Magnet Coil.—If the coil upon one of the magnet limbs should have a partial short circuit, so that the excitation at one pole is greater than at the other, the defect can generally be observed by larger sparks being given at one brush than at the other.

Failure of Dynamo to Excite.—A shunt-wound dynamo will not start or excite upon low resistance. If the binding screws be connected by a short, thick wire, the dynamo will not give a current at all. Similarly, if a line of arc lamps be inserted in the circuit, with their carbons touching, the dynamo will generally refuse to "build up" or fully excite itself to light any of the lamps. For this purpose a resistance coil is frequently inserted in the lamp, which, when the dynamo has fully excited itself is automatically cut out.

Failure to act or excite may be due to the residual

magnetism being too weak. If a series dynamo will not excite when the terminals are connected with a short piece of wire, and the residual magnetism is strong enough, the fault will generally be found in the neighbourhood of the commutator. The brush contact may be bad. The brushes may be partially or wholly short-circuited. The binding screws may be loose, or may be oxidised so as to impede the generation of a current. If the commutator be of the earlier pattern, insulated between its segments with asbestos, it may prove that, pressed hard into the surface of the asbestos, will be discovered a layer of copper dust, or charred (carbonised conductive) oil. Such a cause of short-circuiting of the commutator was once very common, and even now occurs occasionally. In a case of doubt it may be as well to cut out a portion of the asbestos between each pair of segments, so as to expose a clear line of the substance, free from foreign particles.

The leading causes of failure to act in a dynamo are thus short-circuiting or bad contacts.

Repairs to the Armature.

The armature being the moving portion of the machine is more likely to meet with damage than the field magnets or other parts.

Loose Binding.—Taking the case of a drum armature. The wires are generally protected from the effects of centrifugal force, and from being thrown into contact with the ironwork in the bore of the magnet, by binding pieces of steel or brass. This binding is very generally secured by means of tin solder. The tin solder holds very well for a time, but continuous heatings and coolings gradually weaken it and the binding is apt to come off or lose its effect upon the coils. From this cause, and others, the armature wire may come into contact with the magnet when moving at a high rate of speed, and so cause short-circuiting or weakening of the current.

There are numerous other causes of faults in the outer envelope of the armature, such as substances falling between it and the magnet, mechanical injuries by careless handling, and so on. In most cases such external faults can be easily rectified. If a wire is laid bare let it be lifted by means of a bone chisel, and, after being treated with a coating of shellac varnish, wound closely around with silk tape, covered by another coating of varnish. After repair the wire must be pressed quite into its original position and varnished a third time to give it adhesion to the adjacent wires. If the binding wire be loose it must be taken off and replaced by fresh, taking particular care in re-soldering that no drop of the molten metal be allowed to fall between the wires.

A broken wire which is usually too short to reconnect by making a joint is most effectively treated as follows: Strip both ends at the break and clean by scraping. Tin them lightly with the copper bit. Cut an inch of brass or copper tubing large enough to slip tightly over the ends, moisten the interior of the tube with soldering fluid, place the two ends therein, and with a drop of solder upon the soldering bit, fuse all together. Clean off and cover carefully with silk tape and varnish.

Repairs of a more extensive nature, such as rewinding, or placing a fresh coil upon a drum armature, are generally intrusted to the dynamo builder. For re-winding it is always best to send the armature to the actual maker of the machine. Repairs to disc armatures, and all such as have bobbins and coils easily removed from their cores or pockets, are more easily carried out. In such cases fresh coils can generally be kept on hand and slipped on when required. In re-winding a coil upon a "ring" armature, the wire is generally carried in a shuttle, and threaded out and in so as to encircle the ring the requisite number of times. *The number of turns* made by the original coil should be accurately observed, and the same gauge of wire used. Each turn must be drawn tight, and proper insulation applied, with plenty of varnish throughout.

A neat wire splice can be made in a coil by scarfing —or splicing—each end, and filing it rather smaller than the body of the wire; tin the faces of the splice, and solder closely together; file off clean, making the joint rather smaller in the middle than at the ends of the splice; bind it round tightly with a single layer of fine brass wire; tin the whole, and clean off. Insulate as before.

Binding an armature, or re-winding a "reel" coil, should be done by placing the armature or coil between the centres of a lathe. In the re-winding of the Edison type of field magnet, and in several other patterns, the lathe is the most suitable means of rotating the part to be coiled.

Wet Dynamo dried by Steam.—In a recent case, when, by a flood, several dynamos were submerged, they were afterwards completely restored to activity by being dried by steam. The dynamos were covered with tarpaulins, and the steam, at high pressure, applied beneath. After several hours of this treatment, it is said the machines were hot and dry enough to very shortly restore the insulation, and did not suffer in any way by their bath.

Hints to Accumulator Attendants.

The dynamo attendant is generally, in small installations, required to take charge of a battery of accumulators. Indeed, in most isolated or private instances of the introduction of the electric light, the dynamo is only run throughout the day, the accumulators serving to maintain the supply during the hours of lighting. Hence, most dynamo attendants are required or expected to know how to start and manage these secondary batteries. The following hints and suggestions, derived from practical experience, may be of service to the reader:—

The best position for accumulators is in front of a large window, where plenty of light can pass through the cells, and where the attendant can pass completely around them. These facilities for examination are soon found to be of the greatest service. The cells should be of glass, say of the E. P. S. type, than which there is no better cell. They should be raised to a convenient height from the floor upon a dry wood bench; if possible, covered with several coats of a good varnish, especially the top.

Insulating the Accumulator is generally effected by placing under the four corners of each cell the little porcelain cups, filled with resin oil, generally sold with secondary batteries. To preserve the insulation, no liquid should be spilt about the bench, everything should be kept clean and dry. When accumulators are put away in dark, dirty basements and cellars, they cannot be expected to work well. If possible, each cell should be raised above the bench upon a slab of thick (pavement) glass, supported at its ends, so as to allow the light to enter below and facilitate insulation.

Starting and Charging an Accumulator.—For 50-volt incandescent lamps not less than 26 cells will be required, arranged in series. This will give an electromotive force of over 50 volts, allowing a margin for loss in leads. Twenty-six cells will be required for one or more lamps, and 50 cells in series will be required for the ordinary 100-volt lamps. For large numbers of lamps more than one battery of cells, connected in parallel, will be required to generate the amount of current called for.*

When a battery is first set up, and the interiors of the cells quite freed from straw and dust by means of a hand-bellows, it should be connected.

The *brown* plates are the positives. The *grey* plates are the negatives; the grey plates are the smaller. In placing them in the cells they should be carefully handled. The plates are put in, of course, alternately —positive, negative, positive, negative, and so on. The negatives should project equally upon each side so that the positives may be firmly held by the rubber plugs. If the plates have been put in correctly there will be a disconnected positive lug at one end and a disconnected negative lug at the other—of each cell. The cells are connected together, positive to negative, throughout, leaving two opposite lugs, one at either end of the battery. If two batteries have to be used the two positives and the two negatives are to be connected together, making two batteries of an equal number of cells working in parallel.

The *positive* (or brown plate) terminal—generally painted *red*—is intended to be connected to the posi-

* See also "Reserve Cells, p. 51.

tive pole of the dynamo; the negative (or grey plate) terminal—generally painted *black*—to the negative pole of the dynamo.

It is assumed that the attendant understands that the accumulator or storage battery is only of use as a reservoir, or magazine, for storing up the work of the dynamo for use while the dynamo is not running, and that it must be charged and discharged alternately.

Charging.—Before charging the accumulators, if it be in a new station and the capabilities of the dynamo and engine have not been tested, it will be necessary to ascertain both. The attendant should be very sure, by means of a preliminary run of at least a day, that the machinery is to be depended upon before attempting to charge the accumulator. It must not be charged partially and then left for a time. Such a course leads to the rapid destruction of the plates by an action known as sulphating. For the E. P. S. accumulators a run of 36 hours is generally considered requisite, without cessation, upon first charging.

Before connecting the dynamo to the accumulator it may be advisable to test the direction of the current according to the rule given at p. 25.

Do not place the acid in the cells until the last moment before connecting to the dynamo.

The solution is made up by pouring a good quality sulphuric acid into pure water until a specific gravity of 1·170 is shown by the Twaddle hydrometer after proper admixture. The solution should stand to cool before being placed in the cells. Each cell is filled until the plates are covered to the extent of half an inch. Each contact should be examined and tightened up before starting.

CHARGING ACCUMULATORS. 45

The dynamo should either be shunt-wound or separately excited. It must give an E. M. F. of 2·5 volts per cell—say at least 60 volts for a 26-cell accumulator. The first run should not on any account be for less than 12 hours. An automatic cut-out must be used to obviate a back rush of current from the battery if the dynamo should, by any reason, cease working.

The charging must commence immediately after the solution is placed in the cells. If it be delayed, sulphating, or the transformation of the lead plates into lead sulphate, will set in. The same will occur if the first charge be only for a short time. The cells should never, and cannot without certain loss, be left only partially charged and idle.

If it be possible, let the dynamo work upon the battery until the charging is complete, which is indicated by the *milky* appearance of the solution. A great deal of gas, in bubbles, also is given off by the cells before the charging is complete. The bubbles of hydrogen are large, rise into the air and burst, wetting everything near with spray. The oxygen bubbles are smaller and less harmful. Plates of glass are very useful to place over the cells while approaching full charge, but the terminals, or connecting lugs, must be wiped free from moisture occasionally. The moisture will of course collect more copiously under the glass.

When the charging is complete the solution will not only look white, but the hydrometer will show its specific gravity to be at least 1·195. This is a sure test of a full charge. In this, as in every test affecting the charge of the battery, the small voltmeter described at p. 59, should be used.

In subsequent charging make sure of the following :—That the dynamo is running and is excited (its field magnet circuit closed) before switching on the battery. The battery must never be fully discharged, so that a certain current will be generated by it if not opposed by the stronger current of the dynamo. See that the dynamo is switched off before being stopped.

When the accumulators are switched on to feed the lamps the fall in the store of electricity in its plates can be very accurately noted by means of the hydrometer, the specific gravity falling in direct proportion. The attendant will soon, from experience, learn to fix in his mind the amount of current that has been taken out of the accumulator by means of the hydrometer. He will thus be able to determine very nearly how many hours he must run his dynamo to again fill the cells. The gravity should be taken every time before re-charging.

The rate of discharge is calculated from the number of plates in a cell. It is approximately 4 ampères per positive plate of the size designated "L" (E. P. S. type). Thus, a cell containing 15 plates will discharge at the rate of from 24 to to 30 ampères. The accumulator should never be discharged rapidly or upon short circuit. It deteriorates very rapidly under such treatment. A table of the safe rates of charge and discharge generally accompanies an accumulator.

Neither engine nor dynamo should occupy the same room as the accumulator—the acid spray would prove injurious to machinery.

Working Hints.—Agitate the liquid in the cells occasionally, especially during charging. This will prevent the acid from forming in a layer either above or below, and attacking the plates. A little very dilute ammonia kept in shallow open vessels near the

FAULTS IN ACCUMULATORS. 47

accumulator will obviate the nuisance of the acid spray. Do not approach the accumulator with a naked light while nearly charged—the hydrogen given off is apt to cause an explosion. Keep all shelves, supports, insulators, cells, and connections dry and clean. Make up for loss by evaporation by adding water only. Soft water is better than limey, hard water. If a cell in the battery fails to charge fully and is yet clear while the others are white, cut it out and connect across the gap with a piece of cable, properly connected.

Sulphating may be obviated in a great degree by the use of a soda solution, made up as follows :—To a quart of strong solution of common washing soda add slowly, during agitation, 12 fluid ounces of strong sulphuric acid. This should be added to the cells in the proportion 1 part in 25. Sulphating may be obviated by keeping the battery as fully charged as possible. Do not let it lie for days in a half charged condition. If the cells are to be left for some time without working they will take no harm if first fully charged and the insulation, &c., left in good order. The attendant must always have at his hand the hydrometer, a thermometer, and either the simple volt indicator sent out with accumulators, or a standard voltmeter. With those three instruments he can ascertain beforehand which cell is likely to prove faulty.

A *faulty cell*, as before stated, should be at once removed. It is of considerable importance to be able to detect a weak or failing cell before it has had time to destroy itself. It is necessary to maintain all the cells in the accumulator as exactly alike as possible, for if there should be a weak cell the strong ones on either side will rapidly run it down and even reverse it, charging it the wrong way. By means of the voltmeter the condition of the cells can be noted, and any considerable fault detected, but the *temperature* of the cells is regarded as a far more reliable test. All the cells in a battery heat more or less, both while charging and discharging, but *a faulty cell will be warmer than the others*. It will usually emit a hissing sound, louder than the others. The thermometer should therefore be at hand to test temperature, cell by cell, daily. A faulty cell should be emptied of its liquid by means of a siphon. A yard of rubber piping, filled with water, and pinched at either end until one of the ends has been placed below the liquid of the cell, the other hanging down towards the receiving vessel, will be found a ready form of siphon. The liquid should be filtered; the plates should be washed and examined. If they are bent, straighten them, being careful not to damage the plugging. Damage to cells is generally due to *short-circuiting*. Plugs of the lead oxide may fall out of their places in the plates, and short-circuit one or more of the pairs. The importance of placing the glass cells so that light can pass through them, for observation, cannot be too strongly urged. A plug, if it should fall out of its place, may be removed with a pair of hard-wood tongs, or may be made to fall harmlessly to the bottom by a little pressure. There should be enough space between the plates to ensure loose plugs falling to the bottom without short-circuiting the plates.

By way of instruments and accessories the attendant of a large accumulator should be provided with an ammeter, for showing at any time the rate of discharge and the state of the current generally. A bell-alarm, to ring when the rate of discharge is above the normal, to be kept in the circuit. "Excess indicators" of this kind are now to be obtained commercially. They generally consist of certain strips of metal, so fixed in a stand that, upon the current exceeding the safe limit, the heat evolved in the strips causes them to bend and make contact with an electric bell circuit. The bell is generally put in the circuit of one of the accumulators. The distance between the contacts being adjustable, any unsafe amount can be readily provided for. A *cut-out* may of course be used instead, but there is the disadvantage in this that any excess of current, upon acting upon the cut-out, will extinguish the lights.

An automatic switch, for acting upon the circuit while charging, is a valuable accessory. These act by closing the circuit when the dynamo is supplying a current strong enough to charge the cells; but should the current become weak, or other fault occur, the switch will open the circuit and so prevent the accumulator from reversing the dynamo. It may be pointed out that this form of cut-out is especially recommended for use with a series-wound dynamo.

While both charging and discharging it is generally necessary to vary the number of cells in circuit. This is effected by the switch-board, which, as supplied for accumulator installations, is usually provided with switches for "charging," "discharging," "dynamo in," "dynamo out of circuit," &c. In many installations the accumulator is used in conjunction with the dynamo, or more commonly a portion of it. Thus, if anything happens to the machinery, the reserve cells can be switched in, and the absence of the dynamo not noticed.

Leads and Contacts to the Accumulator and Dynamo.—A great deal of waste often takes place by the use of leading cables too small, but more often by bad contacts. Let the lead be amply large enough, and well insulated. Wherever there is a connection that cannot well be soldered, observe, after screwing down the terminal upon the cable and removal, what *surface of contact* exists between the two. The "pinch spot"

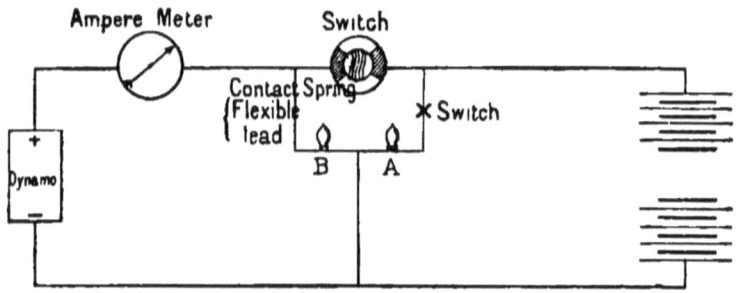

Fig. 10.—Device for Switching Dynamo in Parallel with Accumulator.

will show this. Let the contacts be large, and all such screws protected from oxidation.

Switching-in Dynamos at right instant.—On this point a letter from Mr. Melhuish, of Vienna,[*] describes an ingenious device of his own for overcoming the difficulty of estimating the right time for switching-in the dynamo in parallel with the accumulator. He says, "Perhaps others as well as myself have experienced some little difficulty in putting the charging dynamo on to the accumulator battery exactly at the right time, *i.e.*, just when the E.M.F. of the machine is equal to the E.M.F. of the battery; and I have

[*] *Electrician*, vol. 20, p. 451.

often seen heavy sparking at the commutators and the armatures probably strained by the connection being made too early or too late. Some of the automatic appliances made for this purpose get over the difficulty to some extent, as they are generally set to make the contact when the volts on both sides are approximately equal, but such apparatus are costly at best. I found using a voltmeter also not so convenient as the very simple plan shown diagramatically in the accompanying sketch, Fig. 10. The lamps, A and B, are first selected as giving the same light with the same E.M.F. The lamp A, it will be seen, is lighted from the accumulators, and B is connected with the machine. If now the machine is started B will gradually become bright as the speed increases, and by watching until the light given by the two lamps is equal and then closing the switch, the circuit is made without the least sparking at the commutator of the dynamo, and without throwing any strain whatever upon its armature; for if a double-ended current indicator be placed in the circuit it will be seen that it remains at zero with scarcely a tremor, even when closing the switch."

This plan appears preferable to the use of an automatic instrument, for it is doubtful if it be possible to so adjust an automatic switch as to remain in the same condition month after month, in daily use.

Reserve Cells.—As an accumulator is discharged there is a fall in the potential, but it is so slight that two or three cells, at most, held in reserve suffice to restore the full E.M.F. These also are arranged so that they may be switched in one by one as required. A fall of five volts in a hundred affects the brightness of the lamps.

CHAPTER III.

SWITCH-BOARD AND TESTING WORK.

Running Series or Shunt Dynamos in Parallel.—A good deal of difficulty has been encountered in the running of alternating dynamos (or rather in the switching in) in parallel, especially in public lighting. This has been mostly overcome, however, and it is quite commonly done at all large stations. But the working of constant current machines has been effected in parallel from the first days of electric lighting, probably as early as 1881, and presents few difficulties.

But if a continuous current dynamo is feeding a number of lamps; and if the load upon it begins to be too great, by the switching in of additional lamps, it will not be practicable to merely switch on another machine, even if running at the same speed. Such a course would result in a great electrical strain being put upon both machines, in great overheating, in burning of the commutators and brushes, in dimming or putting out the lights, and, finally, the possible reversal of the fresh machine and the running of it as a motor, if it happen to be the smaller.

The new dynamo must first be "built up," as it is sometimes termed; that is, worked upon an artificial resistance until it is giving a current and pressure at least as great as half the load of lamps upon the working machine.

There are other methods of switching in a new dynamo, but not generally practised, and as the preliminary loading plan is so simple and practicable we will confine our remarks to it alone. In the earlier days of the electric light a bank of lamps, equal at least to half the probable load upon the working dynamo, was provided, suitably connected, and in view of the attendant. The new dynamo was started and worked upon the "bank" until they showed full brightness, compared with a lamp fed by the first machine. The switch was then at once brought into play, and the two dynamos put in parallel upon the main leads. Immediately afterwards the artificial bank of lamps was switched off the new dynamo. Each machine would then take half the load; the first machine would begin to cool down, and the lamps would maintain their brightness.

Artificial resistances in the form of lamps are not now so generally used. Such a course is unnecessary, where the attendant is provided with ammeters and voltmeters. It is more usual to employ either carbon or iron resistance frames. The iron frames are usually either in the form of iron wire wound spirally upon iron tubes, with asbestos separating insulation, or in the shape of zig-zag courses of hoop iron, exposed to the air. It is not of material consequence what form the artificial resistance takes.

Before starting the new dynamos the volts and ampères given by the working machine must be observed. This is usually done by keeping the instruments in the circuit, so that the load upon the machine can be observed at any time. Similar indicators must also be placed in the circuit of the fresh machine, and its working continued upon the artificial resistance until the volts and ampères given by it agree approximately with those upon the working dynamo. The two machines are then put in parallel, and the artificial load at once taken off, as before explained.

For small installations, not provided with many instruments, the simple method of getting the potentials equal before switching in, explained at p. 50 as applicable to accumulators, will be found very useful.

To prevent overloading many stations are furnished with current registers, so arranged that, after the manner of an "excess alarm," a bell begins to ring as soon as the current becomes abnormal. These can be adjusted beforehand to go off at any predetermined load, and obviate any overheating or injury to the working machine.

Alternating Current Dynamos in Parallel.—To effect the running of alternators in parallel without injury to the lamps, or other inconvenience, is not quite so simple as the working of uni-direction machines together.

It may be well to explain that alternators work according to a "phase," a given predetermined number of which are completed in a minute or second, as the case may be. The phase or wave is frequently symbolised by a curved line, ∾ . The number of these per second is known as the rate of alternations, or the periodicity of the dynamo. In Europe the makers of alternators run their machines rather slower than American makers; thus many European alternating dynamos give only 80 to 100 alternations or waves per second, while Westinghouse's American alternator gives as high as 267 per second.

PERIODICITY OF ALTERNATORS. 55

Synchronising, then, of the phases of two dynamos to work in parallel cannot be effected unless they are of the same period and moving at the same speed.

It is generally known that it is easier to work alternators in parallel as their rate of alternations is slower, or, in other words, the switching on is more simply effected, with less liability to disturb the lamps.

If a fresh alternating dynamo, moving at the same speed, and giving the same rate of alternations, be switched at any time into the working circuit of another, a violent jumping of the light, with possible extinctions and re-lightings, will generally occur. This will go on until the two dynamos have worked themselves into the same speed, and are in perfect synchronism. It will be observed that they quickly pull each other into unison.

Such an occurrence would not only be unsuitable to public lighting, but would *rapidly destroy the lamps*. The life of a lamp would, under this treatment, be reduced very considerably, and many lamps, failing to stand the strains, would break altogether. This is not, of course, the only objection to the indiscriminate throwing on of a fresh alternator to a working circuit. The insulation of the machines themselves would probably suffer just as much as the lamps, and there are other objections. Hence the only practicable method is to determine beforehand that the two machines are in perfect accord. Switching on is then attended with no disturbance in the working circuit.

Practically the first dynamo is simply run until its volts and ampères agree with those of the working dynamo. When *perfect* accord occurs the connections are made *instantly*. If well done scarcely a flicker in the lamps will ensue. It is generally effected by means of both instrument observations and one incandescent lamp. A small auxiliary switch places the fresh dynamo parallel with the working machine through one lamp only. This lamp is carefully watched. It usually flickers or dims and brightens frequently before a perfectly steady moment occurs. As soon as the lamp is at full brightness, and is quite steady, the main switch is at once thrown in, bringing both machines upon the same main. If there be any tendency of one alternator to fall out of phase with the other more work will be thrown upon the faster engine, and a balance of half and half is thus almost instantly obtained.

Accumulators and Dynamos in Parallel.—Storage batteries are, of course, only used in connection with continuous current machines. They are being largely employed in central station work, and there has not been found any difficulty in operating them upon mains and feeders in conjunction with dynamos. But in balancing an accumulator to feed in parallel with a machine the use of a rheostat, or resistance, is not required for the accumulator. It is usual to switch in cell after cell until a balance of power is obtained. Again, the simple test lamp device, described at p. 50, may be employed for indicating in place of more elaborate instruments.

Central Station Time and Current Curves.—The use of these is becoming very common. At each central station is kept a table showing at a glance the probable consumption of current at any hour in any given month of the year. This arrangement has led to such a system that the switching attendant can tell almost to a few minutes when to look for an abnormal current from his working dynamos, indicating the necessity for a fresh machine in parallel.

Working Indicators for the Switch Room.

It is curious that scarcely any of the instruments that have been hitherto employed for measuring purposes in the laboratory or class-room have been found useful for the practical work of the dynamo room. This fact has given rise to the production or invention of a goodly number of indicators especially designed for the practical working of installations or systems of public supply, and as most of them are new, or have only come into use within the past year or two, we propose to acquaint the reader with a sketch of two or three of them in the hope that this preliminary information may prove useful to him in employing the instruments themselves. We can only, however, notice a few of the most successful within the limits at our disposal.

Cardew's Patent Voltmeter depends for its action on the expansion of a high resistance wire due to the heat produced by the passage of a current, and is, therefore, absolutely free from the errors due to neighbouring currents and other causes which sometimes exist in other forms which depend upon magnetism for their action. It is said to be the only voltmeter that is self-compensating for temperature and will give the same reading in summer and winter. So far, the Cardew instrument has been chiefly used for alternating currents, but is said to be quite as effective under continuous currents.

The external appearance of the instrument may be observed upon price lists, and merely shows a long tube, carrying at one end a large dial, with a light

58   SWITCH-BOARD AND TESTING WORK.

finger moving over a circular scale graduated to volts and fractions thereof.

The interior of the instruments, as made according to the "1890 pattern," is shown in Fig. 11, where the general arrangement for magnifying the movement due to the expansion of the wire is exhibited. It consists of a series of wheels, or bearer pulleys, usually of bone, set in jewelled centres, with a hair-spring for eliminating "back-lash." The fine wire is carried from the small central pulley away on both sides to the bottom of the tube and back again, so that although the expansion obtained is that due to only half the wire in use the mechanical strain on the wire is halved.

Fig. 11.—Cardew's Voltmeter.

A safety fuse is inserted to save the working wire in case of an excessive E.M.F. being applied to the terminals, but care must be taken in using the instrument not to lift the brushes of a dynamo or produce a sudden difference of potential which might

ACCUMULATOR VOLTMETER.

destroy the working wire before the fuse could act. The wonderful accuracy of this instrument, within certain useful ranges, is becoming well known. It is made generally in two patterns, vertical and horizontal, in sizes ranging from 10 to 30 volts, and similar ranges, three other sizes up to from 40 to 150 volts. The

Fig. 12.—Accumulator Voltmeter.

horizontal tube pattern instruments read up to 120 volts. Messrs. Drake & Gorham are the owners of the patent.

Cardew's Accumulator Voltmeter.—An accurate voltmeter, reading from 0 to 2·5 volts has hitherto been a very difficult one to obtain. In the Cardew cell tester, which will actually read within these figures, we have an accessory indispensable to the user of storage batteries. At p. 47 we gave reasons for the extreme care with which accumulators should be tested for condition, *cell by cell*, at frequent intervals. Although the thermometer will give timely notice of any short circuit in the cell, or other cause that might give rise to heat, there is nothing so certain as a test of the E.M.F. of the cell. The instrument, Fig. 12, is extremely simple, depending as it does upon the slight expansion of a highly resisting wire in two parts, strung in a tense condition between two insulating horns. The wire controls the movement of a pointer so pivoted as to magnify the expansion. The little

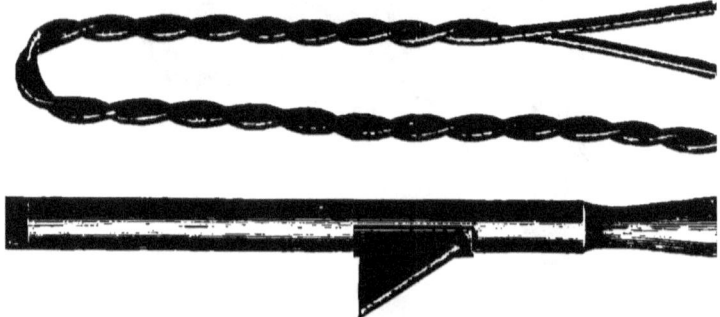

Fig. 13.—Contact Staff for Accumulator.

instrument shown in the figure measures only $3\tfrac{1}{2}$in. square. A careful attendant upon accumulators will use his voltmeter each evening, towards the end of the run, while the accumulator is still discharging upon the lamps, when the lower E.M.F. of any weakly cell can be the more easily detected. But in addition to the use of the voltmeter for detecting incipient faults in the cells it is the proper indicator of the state of the battery while charging. When the cells attain to an E.M.F. of 2·5 volts they are *very nearly* fully charged, and in discharging they should

ACCUMULATOR HYDROMETER. 61

never be allowed to fall below 0·5 volt for reasons explained in the remarks upon accumulators, p. 46.

For making connection with the poles of the cells a useful form of contact maker, Fig. 13, is issued with the small voltmeter. It is only necessary to touch, for a moment, the two poles.

Accumulator Hydrometer Instruments.—While on the subject of instruments for testing accumulators note may be made of the kinds of hydrometers commonly employed by attendants of storage batteries. Fig. 14 represents the usual form of open scale hydrometer with a flattened bulb. It scales from 1·075 to 1·300, an ample range for accumulator work, according to its height floating in the solution. The later form of "bead" hydrometers, Figs. 15 and 16, contain four coloured glass beads, which float at the following densities respectively 1·1050, 1·170, 1·190, and 1·200. These are, of course, more easily read in a poor light than the scale instruments.

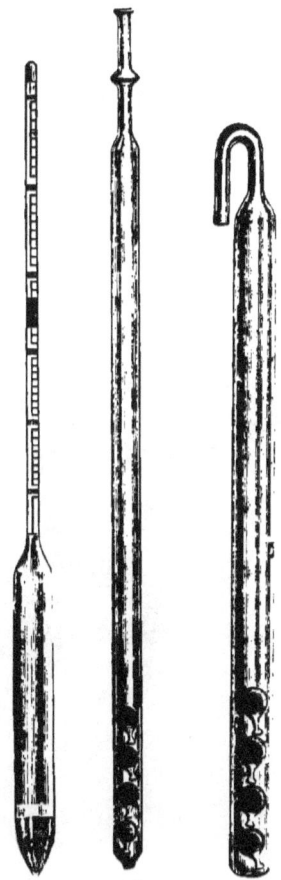

Fig. 14.   Fig. 15.   Fig. 16.
Hydrometers.

Fig. 15 is made much longer, for use in storage cells contained in teak boxes, as used aboard ship, and is known as the ship hydrometer. The instrument is adapted to be passed through the vent hole in the cover of the box, and to withdraw a sufficient quantity of the solution to enable the beads to float. In order that the liquid may not escape, the finger is placed over the opening in the top of the tube. This form is useful also in indicating the level of the solution in the cell. The "Holden" type hydrometer, represented in Fig. 17, as in use in a glass cell, is also largely employed on account of the ease with which its indication can be read. The scale as shown is separate, and set with its point touching the liquid.

Magnetic Voltmeters. — A considerable number of fairly accurate voltmeters, depending for their action upon a permanent magnet, have been introduced of late. Messrs. Ayrton & Perry's is one of the best of these, and is too well known to call for detailed description. There are, however,

Fig. 17.—Holden's Hydrometer.

in certain situations several objections to permanent magnets, the most forcible of which is doubtless the tendency of such a magnet to change in strength, or weaken with time. Such a fault calls for re-adjustment of the instrument at frequent intervals, with all the trouble of having to make or employ an absolutely

CALIBRATING VOLTMETER. 63

accurate "standard" cell, giving a known voltage. This re-adjusting is generally known as calibrating.

To Calibrate Ayrton & Perry's Voltmeter.—It may be useful to owners of these instruments to possess a ready rule for re-adjustment. In both the simple and commutator instruments the adjustment by which the deflections are rendered direct is made by moving the galvanometric coil from a stronger part of the field into a weaker part, or *vice versâ*. The coil is supported by two screws, and by means of nuts it can be moved as above described. On unscrewing the baseboard the magnet and coil of the instrument are exposed, and the adjustment can then be made. To calibrate the commutator voltmeter turn the commutator to parallel, and send a curent from a standard cell of known E.M.F. through the instrument; a deflection, D, will be obtained.* Pull out the plug of the resistance coil, and a new deflection, D, will be given. If E. is the E.M.F. of the standard cell, then the difference of potential at the terminals of the instruments $= E \frac{D-D}{D}$ volts for deflection D, and 1° gives $E \frac{D-D}{DD}$ in parallel, or $10 E \frac{D-D}{DD}$ volts in series. The adjustment of the coil can be made until the desired value per degree is obtained. Although a Daniell cell, giving as nearly as possible, when in zinc and copper sulphate 1·07 volt, is frequently used for calibrating, the result cannot be accurate reading. It is much more satisfactory to use one of Mr. Latimer Clark's standard cells, the employment of which is becoming common among electrical engineers. The E.M.F. yielded by this little cell is accurately 1·435

* A battery of several such cells is usually employed in calibrating.

volts at ordinary temperature. The Clark's cell should never be allowed to work through any resistance less than 1,000 ohms.

Paterson's Electro-Magnetic Voltmeter.—For ordinary work this voltmeter has proved itself very useful. It is at least free from most of the objections urged against permanent magnet voltmeters, and is said to be constant, calling for no re-adjustment. On the other hand the magnetism is got by the setting up of a current, and although this would only introduce a trifling error in a case of considerable E.M.F., it might give rise to a slight drop of potential in the case of delicate readings. The instrument is handy, and not too high in price, with a register accurate enough for every day use. Fig. 18 represents the external appearance. The usual permanent magnet is replaced by a slender electro-magnet, acting upon an indicator in the usual way. It is made to give ranges of from 0 to 5 volts, from 2 to 50 volts, and so on up to 200 volts. By means of extra resistance coils the range of the instruments can be multiplied by 2, 4 or 8.

Fig. 18.—Electro-Magnetic Voltmeter.

Sir William Thomson's " Steel-yard" Gravity Voltmeter is becoming well known to working electricians. It is one of the simplest and most effective in use, consisting as it does of a high resistance coil, in the form of a cone, within which is suspended a short conical piece of soft iron, balanced upon the short end of the steelyard. The connection of the coil to the poles of an electric source causes an attraction of the stumpy piece of iron further into the coil, and a corresponding movement of the steelyard indicator over its scale. Sir William Thomson's new direct-reading vertical scale voltmeter is likely to have extended application for the finer work of the testing room. The same eminent electrician's electrostatic voltmeters, and his recent Centi., Deci., Deka., &c., ampère balances are likely to meet a demand for highly accurate ampèremeters.

A description of Siemens' Electro Dynamometer is scarcely necessary.* Both that and the standard voltmeter introduced by the Edison-Swan Company, as devised by Messrs. Gimmingham & Fleming, are becoming fairly common in collections of electrical test instruments.†

Pocket Voltmeter.—As illustrating the extreme of portability the pocket voltmeter, represented in Figs. 19 and 20, will prove of interest to electricians whose avocations take them away from the more important instruments, which must be kept in the test room, or, at best, cannot be carried in the pocket like a watch. The little volt-

Fig. 19.—Pocket Voltmeter.

* See "Electric Light," p. 331.
† See a paper read before the Institute of Electrical Engineers, Nov. 25, 1887, for a description of the latter instrument.

meter shown is fitted with a permanent magnet, in the field of which is placed a small galvanometric coil, the terminals of which end in two insulated plug holes at the edge of the case. The indicator axis is carried through the face, and terminates in a light style, moving over a graduated scale. The pocket voltmeter is made to read in ranges, from 0 to 10 volts, and so on, up to 80 volts.

Messrs. Ayrton & Perry's Spring Voltmeter and

Fig. 20.—Pocket Voltmeter.

Ammeter has only very recently come into use, but it is likely to supersede many of the more common forms of electro magnet voltmeters. It consists essentially of a coil of wire, acting as a "sucker" coil, the core being a tube of soft iron. To the tube is attached the lower end of a kind of constant diameter volute spring, which carries a pointer on its upper end. Upon the coil being connected to the poles to be tested, the coil excites a down-pulling force upon the soft iron tube, thereby sensibly extending the spring, and causing its upper end to rotate a certain distance, carrying the pointer. It is a very useful instrument for practical work, and acts either as a volt or an ampère meter, according to the resistance of the coil used.

Ampèremeters for Station Work.—Most of the instruments already spoken of as adopted for the measurement of E. M. F. (voltmeters) can also be used, with slight alteration, for measuring current. As a rule, an ammeter is simply a voltmeter with a coil of lower resistance. At many central stations rough, large ampèremeters are put up, composed of a coil of insulated wire, having a freely-moving core of iron or steel balanced within it. The core carries an indicator, or is attached to the short end of a long steelyard, moving over a scale. The scale is usually a rough one, made to correspond with the movements of the indicator. Such a makeshift instrument has frequently been found to preserve its accuracy longer than many more pretentious ampèremeters. Ammeters for installations or central station work are frequently graduated not to ampères, but to lamps, and are adapted to show the number of lamps in the circuit at a glance. Such a system cannot be recommended. It is well known that lamps vary greatly in their consumption, and any instrument graduated to them can only give a rough approximation to the actual current passing.

Whenever makeshift or doubtful instruments are to be used they should be carefully standardised first by means of the torsion dynamometer of Siemens. The ampèremeter devised by Messrs. Fleming & Gimmingham, now issued by the Edison-

Swan Company, and spoken of at p. 65, is likely to have an extended application. But the introduction of commercial ampèremeters, dating as it does back only a few years, has not yet led to the general adoption of any particular system—each electrician chooses or devises for himself.

Fig. 21.—Edison-Howell Lamp Indicator.

Edison-Howell Lamp Indicator, as used on the commercial circuits in America, is represented in Fig. 21. A magnetic needle is employed as the indicator. It remains at "zero" (that is, in this case, the centre of the scale, indicating the "balancing point" or position of equilibrium when set to the exact voltage required upon that circuit), and is only caused to

EDISON-HOWELL LAMP INDICATOR.

deviate either way by any abnormal fall or rise in the potential or current. In principle it depends upon the great variation in the resistance of carbon as the filament of a lamp, due to any change in the temperature. A rise of temperature is accompanied by a fall of resistance in these lamps. In order to utilise this fact use is made of the principle of the Wheatstone's bridge, as indicated in the diagram (Fig. 22). The circuit to be indicated is made to pass a portion of its current through the incandescent lamp. The amount

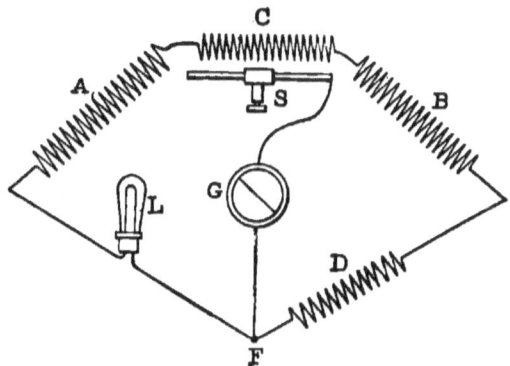

Fig. 22.—Diagram of the Lamp Indicator.

of this current is determined by the E.M.F. or pressure in the circuit, and varies with the pressure. The temperature of the filament varies with the current, as does also its resistance. Thus, by measuring the resistance of the filament (or balancing it) an indirect measure is got of the pressure acting upon it. The marks upon the scale are laid out to indicate the volts necessary to bring the carbon filament to the resistance required, so that the scale gives a direct measure of the E.M.F. in the circuit.

In the diagram L represents a carbon filament

lamp. D is a coil of wire of sufficient size to carry the lamp current without overheating. These two resistances form one arm of the bridge. The other arm is composed of three resistances of wire, A, C, and B, and the galvanometer is connected between the point F and the frame upon which the slide contact S slides. This enables the galvanometer contact to be made at any point of the resistance C. If A, plus that part of C on the left hand side of the galvanometer contact be represented by A^1; and B, plus the part of C on the right hand side of the galvanometer contact be represented by B^1, then when a balance is obtained on the bridge, and no current flows through G, we have $A^1 D = B^1 L$. G is a simple form of galvanometer which indicates the direction and relative amount of the current flowing through it. When the instrument indicator is in zero, no current is flowing in G. If, now, the sliding contact be in the centre of resistance C, and a suitable lamp be in L, on sending a current through the indicator, and gradually increasing it, the resistance of the filament becomes less and less, until, when the desired pressure is acting upon the indicator $A^1 D = B^1 L$ no current flows in G, and the indicator remains at zero with the circuit either made or broken. This position of the sliding contact is marked D, and, if at any time the contact be placed at this point, and the pressure made such that there is no current in G, it is evident that the indicator shows the same pressure as that employed when it was adjusted at that point. The indicator being thus balanced, if the pressure be increased the pointer will move to one side of zero; if it be decreased, it will take an opposite course.

Should the sliding contact now be moved to some

other point, it will alter the values A^1 and B^1, and the pressure will have to be adjusted until the resistance of L is changed enough to make $A^1 D = B^1 L$ again. This position of S, corresponding with the new pressure, may be marked, and in this way can be ascertained the pressures that will balance the bridge with the contact S in any position on the coil C. Thus the scale should be marked for a range of volts. The pressure of the circuit may thus be set at any point that is on the scale, and changing the pressure on the line until the bridge is balanced and the pointer remains at zero, as before, whether the galvanometer circuit be closed or open.

The interior construction of the instrument shows a galvanometer needle carrying a long, light indicator, the terminal point of which is visible upon the upper scale of Fig. 21. A directing magnet is swung upon an axis, accessible from without the case, so that the indicator can at any time be brought to zero. The sliding contact, the indicator attached to which is visible upon the lower scale of the box, has a handle projecting through the case for setting. The coils have a resistance of 90, 257, and 1,000 ohms respectively, and are composed (as to A and D) of German silver wire, and B, partly of German silver and partly of copper; by a known proportion of these the indicator remains balanced at all ordinary temperatures.

Two lamps are usually sent out with each indicator, marked in red and black, corresponding with the red and black marking of the two scales. The object of this is to afford a means of checking the accuracy of the readings by employing one lamp solely for the purpose of comparison, which is used but seldom, and whose resistance therefore is not liable to variation with time. The other lamp is constantly in use, and any variation of its resistance can be ascertained at any time by comparison with the standard lamp.

If the pressure in the mains rises when the instrument is connected across the wires in use, current will flow across the bridge. This will heat both the carbon filament and the German silver wire, but whilst heat increases the resistance of a metal it diminishes that of the carbon filament; hence the resistance of one pair of arms is increased but that of the other pair is diminished, upsetting the balance and causing the galvanometer indicator to deflect from its position of zero. This test is very delicate, and necessarily so, when we consider that a fall of five volts in a hundred in the working pressure will cause lamps which burn quite brightly at a hundred volts to become very dull. On the other hand, a rise of five per cent. will tend to shorten the life of the lamps. Upon well-conducted systems the pressure upon the mains is never allowed to vary more than one-half per cent. Such close working could not easily be attained by any ordinary means other than the use of zero-indicating instruments such as we have described.

Regulation of the Dynamos to correspond with Balance Indicator.—As we have already ascertained (page 6), the regulation, save in small installations where a compound-wound machine may regulate for itself, is usually effected by varying the exciting current by means of a shunt. The exciting machine field is indeed the point of regulation generally resorted to. It is comparatively easy, by means of a simple resistance frame, to so control this small current that the pres-

RHEOSTATS.

sure is kept within the required half per cent. At some stations the speed of the engine is accelerated or retarded as required.

Enough has been said to give some idea of the nature and variety of indicating and regulating instruments employed in isolated and central station plant working.

A few words further, by way of summing up. The general scope of these indicators and controllers may serve to fix in the reader's mind the everyday application of this lately-developed system of regulating electric supply.

Of *Rheostats* it may be said that for practical dynamo-room work they are simply resistances that can be adjusted from little to much at will. They take many forms. For heavy current work large rheostats usually consist of coiled (or otherwise compactly disposed) iron wire, or of hoop-iron, or of carbon-rod; or, for instrument work, the more expensive German silver wire. The practice of controlling by means of a variable resistance, save for small currents, is going out of fashion. At one time, and even now, in isolated plants, the main current was passed through a resistance which was raised to suit the lamps. This was extremely wasteful, and a great step in advance was taken by controlling only the exciting current of the field magnets—in other words, varying the resistance of the shunt. It is of course still more economical to vary the current of the field of the exciting dynamo when a separate exciter is employed. Dynamo *balancing rheostats* are merely resistances having a range sufficient to balance the current of one dynamo against that of another, with which it is desired to run it in parallel. This was formerly, as explained at

p. 53, effected by a rheostat of lamps, or merely a bank of lamps, not adjustable.

The *regulating* part of the rheostat usually consists of a series of contacts, or copper plates, connected in series with successive portions of the iron wire or hoop. By the sliding of a lever over them each one is in turn switched over into the circuit and adds its resistance thereto. They are usually made in the form of a circle for ease in manipulation. It may be accepted as a sign of poor electrical engineering when large rheostats have to be used at all. Their use means waste and bad regulation.

Résumé.—Broadly, then, the instruments used in the dynamo and switch-rooms are as follow:—Rheostats, for regulating the E. M. F. and current of the dynamos; balancing rheostats; potential indicators (voltmeters) for showing at a glance any variation in the pressure upon the leads or mains (*brightness of the lamps*). These usually remain continually in the circuit. Ampèremeters (current meters) used to indicate the *number of lamps on circuit*. This is also commonly kept in circuit, and shows, in conjunction with the voltmeter, the work going on, giving warning of any necessary approaching change. Detectors, for calling attention to ground or earth faults. These take several forms, according to the nature of the work and the potential at which the mains are charged.

Resistance and Insulation Testing.

The "practical" electricity of the schools, classrooms, and laboratories is frequently so different from the practical electricity of the electric light station that a few hints respecting the taking of resistances appears necessary in the present chapter.

RESISTANCE AND INSULATION TESTING. 75

The resistances that most frequently need to be measured are those of the coils of armatures and field magnets of dynamo machines, the resistances of mains or leading cables, the resistances of branch leads and wiring generally in buildings. These are resistance tests only, but they only give half the information generally required. It is frequently still more necessary to ascertain the *insulation* resistance of the coils of armatures and field magnets, and of mains, feeders, branch leads, and wiring generally.

A rough measurement of the resistance of an armature, or field coil, is sometimes taken as follows. It may, of course, be made sufficiently accurate for ordinary purposes by using sufficient care in taking the test. It depends upon the principle of comparing two deflections of the galvanometer, which are proportional to the fall of the potential shown by the insertion of a resistance.

Take a wire measuring over a hundred ohms, and another wire of only a hundredth of an ohm. These can be obtained from, or produced by, the testing set described further on, if not at hand. Connect these in series with an accumulator cell, and the brushes pressing upon the commutator of the dynamo, forming a complete circuit: cell, two resistances, and armature coil. For the test employ, if possible, a delicate reflecting galvanometer; connect it by wires with the two ends of the one hundredth ohm resistance and note the deflection. Remove the wires and apply them to the brushes of the machine, turning the commutator round, sector by sector, so as to get the average deflection. The average deflection produced upon the galvanometer, compared with the deflection got from the one hundredth ohm coil, show at once the resistance of the armature coil compared with the hundredth ohm coil; *e.g.*, a deflection of 50° from the armature coil and of only 25° from the hundredth ohm coil would show that the armature coil had as much again resistance as the hundredth ohm coil. A "comparison coil" of any suitable resistance can be used, but when the resistance of the armature is low the hundredth part of an ohm is a convenient standard. The rough coils employed for this purpose are frequently of hoop iron. Strap iron, having a thickness of $\frac{1}{32}$nd of an inch and a width of half an inch, will be found to give approximately one ohm resistance for each 100 yards. The high resistance coil is merely used as a "choking coil" to obviate the passage of an appreciable current by the cell. The setting up of a large current, which would ensue upon using a small resistance, would probably entail a fall of potential in the cell, and an error in the observation.

This simple method is frequently used for measuring the resistance of circuits, its chief advantage being that no instruments are required except a reflecting galvanometer, and only one of the resistance coils—the smaller—need be accurately known.

Testing Box.—But for general resistance measuring a proper "testing set," consists of a Wheatstone bridge combined with resistances in a case, and a testing battery in a separate portable form. The nature of the balancing method generally spoken of as Wheatstone's bridge is well known to students, and since it is elucidated in class-rooms, and may be studied in any standard text-book of electricity, it will be unnecessary to describe it fully here. The class-room Wheatstone bridge, however, is not quite similar to that used by practical electricians, and we

WHEATSTONE BRIDGE.

therefore select for description a good form for everyday working purposes.

Fig. 23 represents the general appearance of a testing box, as issued by the Gutta-Percha and Telegraph Works Company, and is especially designed for the use of electric-light engineers.

Fig. 23.—Testing Box.

It consists of a Wheatstone bridge with dial pattern resistance coils, capable of measuring copper resistances of from ·01 ohm to 9,900 ohms, a highly sensitive galvanometer, and all the necessary appliances for measuring insulation resistances up to 30 meg-ohms, a galvanometer key, and $1/_9$th and $1/_{99}$th galvanometer shunts, fitted complete in a teak-

78   SWITCH-BOARD AND TESTING WORK.

wood case, the dimensions being 7½ in. × 7½ in. × 6⅝ in., and weight 8¼ lbs.

Such a testing box saves a great deal of trouble. It obviates the necessity for setting up the delicate and by no means portable instruments sometimes employed where reliable tests are to be made.

The battery consists of 30 Leclanché elements of a special portable pattern, contained in a polished wood box 11 in. × 9 in. and 7 in., connected up to terminals in the box, and provided with flexible wires for connection to the test-box.

To take a Conductor Resistance.—Fig. 24 represents the particular arrangement of the resistances, terminals and galvanometer in the box forming the Wheatstone bridge. Figs. 25 and 26 show the connections necessary in taking copper resistances and insulation tests respectively. It may be pointed out that *insulation testing* is daily becoming of more and more importance. The fire offices now insist upon stringent conditions of safety insulation, and it may be accepted as the tendency of the times to provide abundant "safety" insulation as distinguished from purely electrical insulation, which might in many cases be of a comparatively inferior nature and yet, serve the purposes of carrying the work of feeding lamps on from day to day, without much loss.

Reverting to Fig. 24, A and B are Wheatstone-bridge balance coils, and C and D, two dial-form resistances; G the galvanometer, and K the key.

Connect the conductor or the circuit to be tested to the " bridge terminals " on the right front of the box ; the plugs connected to the battery wires are to be placed in the " bridge " plug holes in the right of the box.

TAKING CONDUCTOR RESISTANCE.

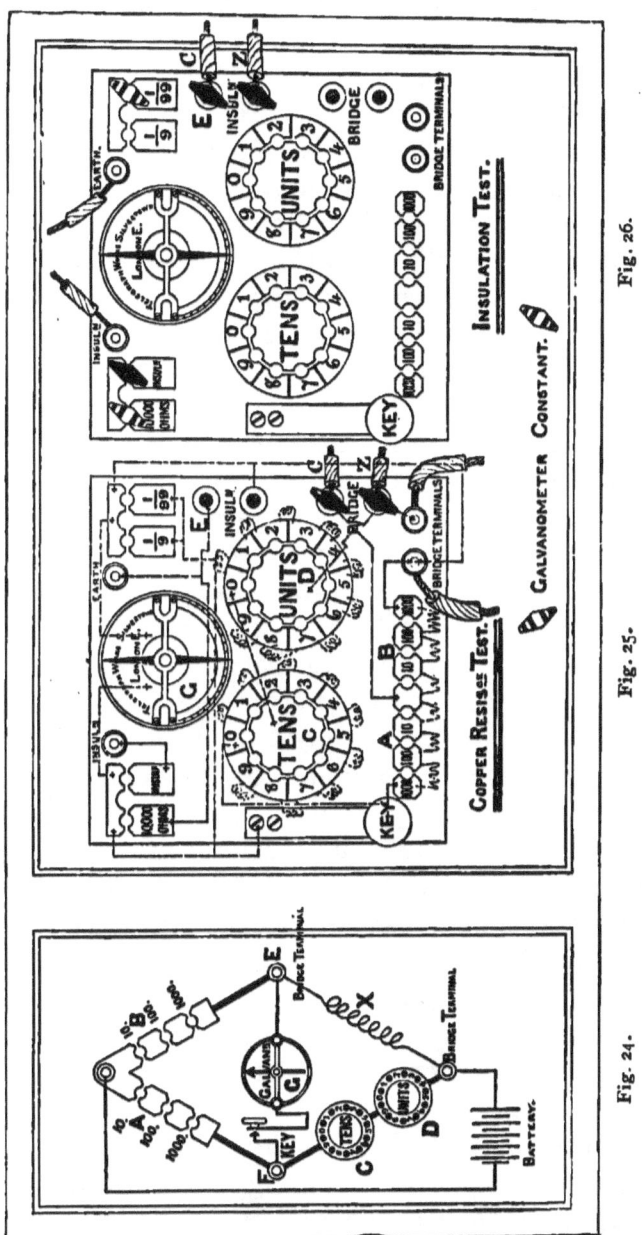

Fig. 24.

Fig. 25.

Fig. 26.

Diagrams of Conductor and Insulation Resistance Tests.

The resistance of a conductor is obtained by balancing known resistances against the resistance to be measured in the following manner :—

The resistances marked A and B are the ratio resistances, and in each test it is necessary that one should be unplugged in A and one in B.

The theory of the arrangement is the obtaining of equilibrium by the adjustment of the resistances in A, B, and C, D, until there is no difference of potential between the points E and F, and consequently no deflection of the galvanometer needle when the key is closed. These conditions can only be obtained when the resistances in the two sides X B and C D A are equal, or bear certain proportions to each other. Let us take the case of obtaining equilibriums with equal resistances. Make the resistance of the ratio sides A and B equal by unplugging the 10, 100, or 1,000 coil in each; it will be obvious that a balance or state of equilibrium between the points E and F will be obtained when C, D=X; it is therefore necessary to vary C, D until no deflection of the galvanometer needle is produced on repeated tapping of the galvanometer key, when C, D=X. It will be observed that by using equal ratio resistances, any resistance between 1 and 99 ohms can be measured, but by a suitable arrangement of the ratio resistances the range can be extended to from ·01 ohm to 9·900 ohms, for if the 10 coil in the ratio arm B, and the 100 coil in the ratio arm A are unplugged, a balance will be obtained when the resistance in C, D is ten times that of X; therefore, C, D divided by 10 will give the resistance of X. In the same manner we may have the 10 coil in B unplugged and the 1,000 coil in A, in which case we divide the resistance in

INSULATION RESISTANCE TESTING. 81

C, D (when a balance is obtained) by 100 to obtain the resistance of X. High resistances are measured in the same manner, but the resistance in ratio arm B is made higher than that in A. For example, if we make B 100 and A 10 we multiply C, D by 10 to obtain X, and if B is 1,000 and A 10 we multiply C, D by 100. In the testing box the ratios are placed in front of the ebonite base; the left hand 1,000, 100, and 10 coils representing A, and the right hand 10, 100, 1,000 coils representing B.

The following table gives the most suitable ratios for measuring resistances between the limits stated :—

Between		Ratio		
Ohms.	Ohms.	Right hand.	Left hand.	
900 and 9,900	..	1000 ..	10	Multiply C, D by 100.
90 ,,	900	.. 100 ..	10	,, C, D by 10.
9 ,,	90	.. 10 ..	10	C, D = X.
·9 ,,	9	.. 10 ..	100	Divide C, D by 10.
·01 ,,	·9	.. 10 ..	1000	,, C, D by 100.

It will be found while adjusting the dial resistances C, D, that if the resistances to which the dials are adjusted is higher than X, the galvanometer needle will be deflected to one side, while if the dial resistance is lower than X, the deflection will be to the opposite side, becoming less and less as the balance is approached. When the balance is nearly obtained, the key should be pressed down repeatedly, in order to induce the galvanometer needle to swing.

To take an Insulation Resistance Test.—To obtain a constant, place the battery-wire plugs in the plug holes marked insulation, and plug up the 10,000 and $1/9$ or $1/99$ shunt in order to obtain a suitable deflection of the galvanometer needle; call this deflection θ, and the shunt used S.

Insulation :—Connect the terminal marked "earth"

G

to any convenient ground contact, such as a gas or water pipe, and that marked insulation to the conductor or circuit to be tested. Plug up the "insulation" switch (removing the plug from 10,000) and, if required, $1/_9$ or $1/_{99}$ shunt, reproducing as nearly as possible the constant deflection θ. Call the deflection of the galvanometer pointer D, and shunt S, then—

Insulation resistance in ohms=
$$\frac{\theta \times S \times 10,000}{D \times S} \quad (1)$$

or, if no shunt has been used in the insulation test—

$$\text{Resistance in ohms} = \frac{\theta \times S \times 10,000}{D} \quad (2)$$

Note.—The multiplying power of the $1/_9$ shunt is 10, and that of the $1/_{99}$ shunt 100.

Example.—Suppose the deflection θ, when taking the constant to be 45°, and the shunt being $1/_{99}$, and the deflection D 20° with $1/_9$ shunt, then, according to equation 1—

$$\frac{45 \times 100 \times 10,000}{20 \times 10} = 225,000 \text{ ohms.}$$

Example.—The constant deflection being as above, let the deflection D = 5°, no shunt being used—

$$\frac{45 \times 100 \times 10,000}{5} = 9,000,000 \text{ ohms or 9 meg-ohms.}$$

Portable Wheatstone's Bridge.—A very convenient portable bridge, for the comparing of small resistances (up to 100 ohms) has lately been introduced by Messrs. Woodhouse & Rawson. It measures only $5\frac{1}{2}$ in. in diameter, with a height of $3\frac{3}{8}$ in. The ratio wire is of platinum silver, arranged in a circular spiral form and stretched upon the thread of a double threaded screw, cut on an ebonite cylinder, and the connecting wire to one terminal of the battery

TESTS DURING WIRING. 83

stretched on the other thread. Connection is made between the two by a spring and roller contact fixed to the inside of the ring-nut working on the screwed cylinder. The position of this nut, with its contact piece, determines the relative lengths of the platinum silver wire on either side. Four plug blocks are placed on the top of the cylinder, allowing a resistance of either ·01, .1, 1, or 10 ohms to be inserted in one arm of the bridge. Here are also the necessary terminals for connection to battery, galvanometer and resistance to be measured with a contact key. The arrangement is represented in Fig. 27.

Insulation and Conductivity Tests during Wiring. — The simplest test, and indeed that to which the common wireman confines himself, is a test for continuity. The

Fig. 27.—Portable Wheatstone's Bridge.

only instruments required for this consist of a battery cell of any portable kind, and a galvanometer of the simplest description. A wireman's "testing set" usually consists of a semi-dry Leclanché cell, a detector galvanometer, having a few hundred ohms resistance, and a connecting key, with suitable terminals. The whole is usually fitted into a small portable wooden case. But the only information such an equipment gives is warning of any break of continuity in the wiring.

There may be a serious defect in even the continuity, which the first attempt to pass a large current would make manifest, and yet the continuity test might be satisfactory. Such a fault commonly consists of any one of the following :—

A wire practically broken, but yet partially connected by a portion of the wire only.

A wire broken right across, but touching at the break.

A twisted joint, without being soldered, but yet loose, *e.g.*, not rigid.

A loose safety plug or cut out, and yet in contact sufficient to give a deflection upon the galvanometer.

Wires are sometimes accidentally caught by the wireman's cutting plyers and cut *nearly* across without his knowledge. Such a point would heat very considerably when carrying full currents, and would finally be burned, opening the way for the formation of a dangerous arc between the severed ends. An arc of this kind would probably produce a fire.

To take the ordinary conductivity test it is only necessary to connect the battery cell, galvanometer, and wire, to be tested in series, and to complete the circuit by connecting the remaining terminal of the battery to the " return " end of the wire. The wire may, of course, be connected at its far end to the " general return lead," and in the case of ship wiring to the iron body of the vessel, but the connection through the wire will be the same as if the wire itself were continued back to the test box. If there is continuity a deflection of the galvanometer will ensue upon depressing the key. If there is a break there will be no deflection. If there be a very weak deflection, it may be due to ineffective connections at the box itself, or to some one of the defects already indicated.

TESTS DURING WIRING. 85

Resistance Tests.—These are taken by several methods, and it may be said that each electrician has his own particular favourite plan. The two leading methods in practical use, however, are that given at p. 78 and known as the balance test (by Wheatstone's bridge), and that described at p. 75 known as the comparison of fall of potential method. The Wheatstone's bridge test is the most usually applicable, save to very small resistances, and will be found all that is necessary in everyday working.

The resistance tests are usually deferred until all the wiring is complete. Each circuit can then be taken in turn. In the case of an "installation" of the electric light, as in a house or ship, the tests should be taken from the main switch-board, close to the dynamo. In the case of house wiring for central station current the tests will begin at the distributing board where the leads join the connections (from the main) leading into the street. Circuits that are previously planned will probably be also estimated for resistance, and it should thus be very nearly known the resistance in ohms presented by each circuit. Such preliminary information facilitates and ensures the correctness of the tests, because the tests themselves should approximate closely to the calculated resistances. If they should vary the cause must be sought out. The tests themselves will be found fully discussed at p. 78.

It scarcely seems necessary to remind the reader that the extremities of the "branch" and "twig" leads must be temporarily connected together during the copper resistance tests; otherwise the movement would be through the carbon filaments of the lamps if these happened to be connected. But lamp connections must be kept open during the tests.

Insulation Resistance Tests.—This test is becoming of greater importance in the case of house and ship wiring than the conductivity test itself. It is the test to which the fire offices will look for safety, and its fulfilment should be insisted upon by every fitter of the electric light.

The insulation test is of particular importance aboard ship and in all buildings where damp is likely to injure the insulation. The apparatus employed should consist of such a testing bridge as that described at p. 78 with its battery of 30 cells, or as many more cells as can be conveniently used. A small battery power is useless. The test itself is most easily taken according to the method given at p. 81. It should show an insulation resistance of at least ·1 meg-ohm * per lamp for every volt to be used on the circuit.

The earth connection made in taking the test should be particularly good. A water-pipe is usually a good earth—better than a gas-pipe, where joints interfere with the conductivity to earth.

The insulation aboard-ship, in cases where "ship return" is used, should be greater than that given above. In this case the "earth" will be the shell of the ship.

But an insulation test may show full value before the working current is passed, and may fall off under the E.M.F. of the dynamo. For this reason the test should be in duplicate before the current is turned on to the lamps, and after the lamps have been run for many hours. Aboard-ship, and in damp situations, the test should be repeated at intervals.

* Meg-ohm = one million ohms.

Insulation tests of overhead wires will show high in dry weather and low in damp weather. In the case of naked mains, underground, the same will hold good. Insulation tests of dynamo and transformer coils should be at least as high as that cited.

Details relating to installation and house wiring testing are more fully treated in Chap. V, p. 182.

CHAPTER IV.

ARC LIGHT WIRING AND FITTING.

A GREAT deal of the trouble that has hitherto been encountered in the general utilisation of arc lamps has arisen from ignorance on the part of those fitting up the circuits for, or attending to, the arc lamps. Now that the lamps and currents are becoming better known, more skilled attention is given to them, and a great impetus has thereby been imparted to arc lighting. It may also be pointed out that the revival of arc lighting (which, until lately, appeared likely to be eclipsed by incandescent lighting) is due in a considerable degree to improvements in the lamps themselves, to more effective insulators, superior methods of automatic regulation at the dynamos, and to a better quality of carbons.

The wonderful cheapness of the arc light, compared to the incandescent light, would prove a strong inducement to many to adopt it in place of gas, were it practicable to obtain in ordinary commerce plant that could be depended upon to yield a *reliable* and *steady* light. The general opinion of the arc light was, until very lately, that it had not yet passed the experimental stage, and that it was in consequence erratic and unreliable.

It should now be the aim of everyone interested in the new light to remove that impression, and to assist

ARC LAMP COILS.

in spreading a general conviction that arc lighting can be depended upon; that it is inexpensive and safe.

This can only be effected by first understanding the nature of the arc-lighting current, the nature of the dynamo producing it, the best means of controlling it, the most effective method of distributing the light, and the working and care of the arc lamp itself. In one item alone (carbons) the cost of the running of the light has recently been very greatly reduced. And by such improvements as the substitution of a laminated for a solid armature in a dynamo the cost of the light has been reduced by one-third.

The Obsolete Single-Arc System.—When arc lamps run from dynamos were first brought into use only one lamp could be put into the circuit of one machine. This was justly considered a great bar to the diffusion of the light, and altogether a costly system, necessitating as it did powerful lamps, only suited to special restricted areas. When Jablochkoff introduced his electric candle it was thought that arc lamps were entirely superseded, since several candles could be burned in the circuit of one dynamo. But an improvement was soon effected in the arc lamp, which opened up a means of putting any number into the circuit of one machine. We may glance at the early form and the improved form. .

Single-Regulating Coil Arc.—All the earlier lamps were regulated by means of a coarse wire solinoid, through which the whole of the current passed. If the arc in this lamp became too long the solinoid, through the weakening of the current, would allow the carbon rod to drop, producing a sudden "wink" and re-establishing the proper length of the arc. This sudden disturbance *on the main and only circuit* would be communicated to any other lamp on that wire, and would upset its arc. Every movement of the solinoid core, or every defect in the arc, would thus necessarily be imparted to the whole circuit. It was therefore impracticable to run more than one such lamp on one circuit. If two or more were put in a constantly flickering light ensued; some lamp would always be adjusting its arc and disturbing the others. This fault was overcome by a marked improvement, called a—

Differential or Shunt-regulating Coil.—The coarse wire coil was retained as before, with its iron core, supporting the upper carbon; but wedded thereto was a fine wire coil connected as a shunt to the arc. That is, when the current reached the positive terminal it had two paths open to it—through the coarse coil and the arc, and through the shunt coil direct back to the machine or negative terminal. If the arc became too long the current *through it* would tend to weaken, but this would cause a correspondingly stronger current to flow through the shunt, so that the current in the wires outside this lamp was not weakened. If the arc became too short the current *through it* would become stronger, but this would cause a correspondingly weaker current to pass through the shunt coil, so that the current on the outside wires was not strengthened. The shunt coil also exerts a control over the separation of the carbons, and by means of these *constantly balancing* factors, the current passing through the lamp is practically constant, and does not affect any other lamp.

These elementary explanations are offered in order to clearly distinguish in the reader's mind the nature of the lamps, with their balancing devices, suitable for single and multiple lighting. The construction of the lamps themselves, in which many ingenious modifications have been introduced, can be studied in a good descriptive book on electric lighting.* The modifications are chiefly in the direction of making the lamps suitable for either *constant current or constant potential*. There is, at least, one lamp (the Brockie-Pell) which is so ingeniously adjusted as to fit it for working upon either circuit—that is, it may be taken from a constant current circuit and placed in a circuit where the potential is kept constant instead of the current, and it will burn very well. There is also a successful automatic cut-out in each lamp, so that if the carbons happen to burn out, the circuit will not be interrupted, but will remain open through the cut-out, leaving the lamp in a by-pass.

Arc lamps in parallel, that is, placed across the main leads, like incandescent lamps, work very well when fitted with resistances or "choking coils," as they are called.†

Arc lamps are frequently run upon *alternating current circuits*, when adopted for that purpose, and in this way are now run off transformers along with incandescent lamps—in this case, the arc lamp is, of course, placed in parallel across the leads.

Focussing arc lamps are those in which both carbons move towards the arc, are burned equally, and keep the arc in one place.

The distance between the carbons in arc lamps does not vary much. It always depends upon, and is nearly proportional to, the E.M.F. in the circuit.

With an E.M.F. of 50 volts and a current of 15 ampères the usual working distance is $\frac{3}{16}$ths of an inch. With a 40 volts E.M.F. and a 10 ampère

---

* See pp. 206—250 of "Electric Light," 4th ed. London, 1890.
† " Choking and Impedance Coils, p. 115.

current, $\frac{1}{8}$th of an inch is usually the most effective working distance. These figures apply to the ordinary powerful arc lamps used for street lighting when worked *in series;* when worked in parallel, the E.M.F. must be higher—usually about 20 per cent.

When a dynamo is running only one lamp its E.M.F. need not exceed 50 volts.

When a dynamo is running several lamps its E.M.F. must be proportional to the number of lamps. If each lamp calls for 50 volts, and there are 20 lamps *in series*, the E.M.F. developed by the machine must be at least 1,000 volts, with an allowance for fall of potential due to the leads and branches.

When a dynamo is running lamps in parallel its E.M.F. need only be high enough to run one lamp, with the usual allowance for resistance of leads. But its *out-put of current* must then be in proportion to the number of lamps; if one lamp takes a current of 10 ampères, and there are 20 lamps, the current must be at least 200 ampères.

The Series method of running is more economical.

Arc lamps are in use—*e.g.*, Siemens' differential ; the Brokie-Pell—capable of giving a steady light with a current of only 4½ ampères, in series. These are chiefly employed for indoor lighting. The minimum E.M.F. to secure a clear arc is probably 30 volts—*e.g.*, the carbons will not separate, and produce the true arc with less.

Regulation when running Arc Lamps in Series.—An ampèremeter is always placed upon the circuit near to the dynamo, so that the attendant can see at a glance the current flowing. He is chiefly concerned in keeping the current constant. This is frequently done by switching in more or less resistance; into the main circuit, if the dynamo be a series-wound one, and into the field magnet circuit if it be a shunt-wound machine; but the shunt or compound wound machines are supposed to regulate themselves, which they very often fail to do. Changing the speed of the engine is more generally applicable to the regulation of constant potential circuits. There are several automatic constant current regulators, in use more or less efficient.

Regulation when running Arc Lamps in Parallel.— A voltmeter showing the volts is in constant use across the leads, and under the eye of the attendant. His chief care is to keep the *potential difference* between the leads the same. This is usually effected in part by the dynamo itself, when a shunt-wound machine is used, or by regulating the *speed*, which has in most cases a direct control over the E.M.F.

Arc Lamp Trimming.

Unsteadiness of the light is usually caused by small defects that are allowed to develop through the attendant not understanding his lamps. To work a lamp to the best advantage, especially if it be out of doors, exposed to wind and rain, calls for some little skill and familiarity with the mechanism of the lamp.

When lamps burn unsatisfactorily, and the cause cannot be found in the regulating mechanism, it may be due to the carbons used being faulty, or to poor insulation of the leading wires, but more frequently to the current or pressure (E.M.F.) not being suitable to that particular lamp. To obviate this the maker of the lamp should always issue with it the necessary particulars of pressure and current at which it is intended to burn. But most arc lamps contain within

94 ARC LIGHT WIRING AND FITTING.

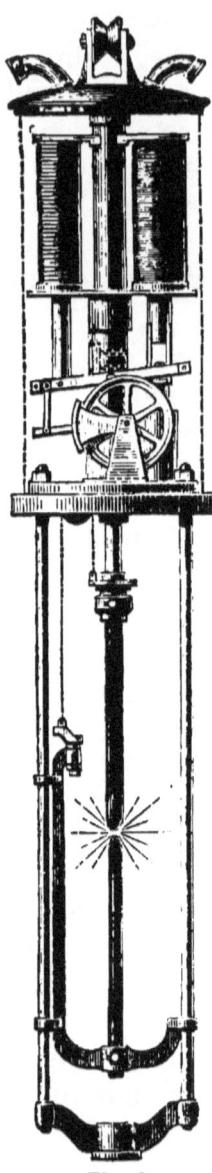

Fig. 28.
Brockie-Pell Lamp.

themselves mechanical balances, or other devices to enable the user to regulate them for himself, and to suit the pressure and current given by his dynamo. The carbons used must always correspond with the current, *e.g.*, thick carbons for a large current and thin ones for a small current. The length of the arc is always regulated to suit the pressure.

Fig. 28 represents the working arrangements of the Brockie-Pell lamp, and Fig. 29 its external appearance, encased ready for work.*

As an example of the attention that must be given to arc lamps to run them successfully, we append a few working directions applicable to the Brockie-Pell Arc Lamp.—

1. This lamp is regulated for a normal current of — ampères.
2. The current used must not exceed —.
3. The current must not be less than —.
4. To work at the maximum current (— ampères) take off the lead weight on top of piston of dash-pot.
5. To work with the minimum current (— ampères) fill the hollow piston with small shot.
6. In fact, to work with small

* For details see p. 218, "Electric Light," 4th ed.

ARC LAMP TRIMMING.

currents, *add* weight to piston; to work with large currents *subtract* weight.

7. The piston is easily removed by simply unscrewing the dash-pot from the base-plate of the lamp; the plug at the top of piston unscrews, and the shot can then be added or taken out. Take care not to bend the piston link in unscrewing the plug, which may have become set fast. Put *no oil* on the working pivots or any other part of the lamp. *On no account put liquid in the dash-pot;* this should be absolutely dry, and will keep in good working order for many months; when it becomes too stiff simply wipe it out; use no emery or other cleaning powder in this or any other part of the lamp.

8. Before inserting new carbons always wipe the rack, or sliding-rod and guide-rods, with a piece of soft leather; attention to this simple rule will keep the lamp in good order, whilst neglect will probably soon cause the rack to stick.

9. After the lamp is hung up, or fixed in position, see that the guide-rods have not become twisted; if they have set them perpendicular with the centre-rack.

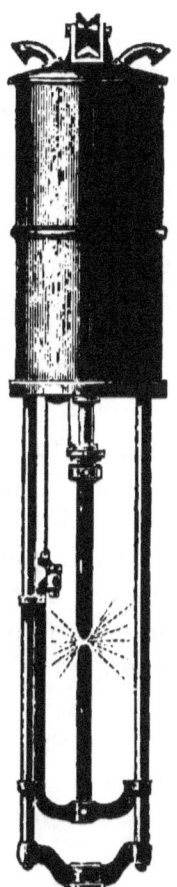

Fig. 29.
Brockie-Pell Lamp (encased).

10. To get the carbons exactly *in line*, and their points central with each other, insert the top *cored* carbon first, clamp it in the holder, and then observe if the point of the carbon naturally points to the centre

of the lower carbon holder; if not, unclamp the carbon, turn it round a little, and try again and again until it points fairly toward the centre, then insert the lower *solid* carbon, and adjust its point to the upper.

11. To secure a good light the distance between the carbons when burning should not exceed $\frac{3}{16}$ths of an inch, and $\frac{1}{8}$th of an inch is quite enough for a 10-ampère current.

12. The wires from the machines must be connected to the terminals of the lamp that the upper, or positive, carbon burns hollow and much brighter than the negative, or lower, carbon, which, on the other hand, burns to a slight point.

13. The rule is to connect the positive wire of the machine to the uninsulated terminal of the lamp. Should the carbons burn so that the light is thrown upwards, the wires must be reversed at the machine. If the lamp goes out and relights itself frequently, or a pumping action of the regulator takes place, it is a sure sign of the current being too weak, or of the machine magnetism being unstable—the lamp will *never* have this pumping action unless the current is at fault.

14. Should the lamp short-circuit itself by means of its automatic cut-out, the hand-switch must be used for relighting the lamp, unless it relights automatically.

15. The electromotive force necessary for each lamp when in series is about 45 volts; if the lamps are worked in parallel the electromotive force must be at least 55 volts, and a resistance must be placed in each lamp circuit to reduce the current in the lamp to what is required.

16. For parallel working on a — volt circuit this

lamp requires — ohms resistance in series with the lamp.

Adjustment of Brush System Lamp.—The treatment of arc lamps in which glycerine is used, in either dash-pot or carbon-holder rods, or in both, is rather different from the foregoing. The "dash-pot" in an arc lamp is simply a small cylinder, fitted with a piston and piston rod, the function of which is to prevent jerky or sudden descent of the carbon rod. In many lamps merely the piston, acting upon the air within the cylinder, is employed. In others glycerine, or a mixture of glycerine and water, is used to modify the movements. In the Brush lamp the dash-pot has to be unhooked from the armature and unpinned at the top, and then half filled with a mixture of three parts of glycerine and one of water. There is an air-hole in the top which must be kept clear. In replacing the dash-pot it must be so adjusted in its position that when raised to its highest point it is quite free from binding. The pin which secures the upper end of the piston rod must pass easily into its place, and must not bind in the hole in the rod itself. The piston rod must pass quite freely through the cover of the cylinder.

The brass carbon rods are tubes which carry the upper carbons (in twin-carbon lamps) and are the most important parts of the lamps as regards cleanliness, the perfect working of the lamp depending upon the regularity with which they feed the carbons. They may work irregularly through any foreign matter being attached to them—oxidation, or gummed oil. The fault may be in the bushes through which they pass, or in the tilting washers or rings which raise them, or in the pistons within them which

H

govern their fall, and prevent it from being too rapid.

In cleaning carbon-holder rods in arc lamps any polishing powder, as bath brick or emery, must never be used. This rule will be found to apply to all these lamps, embracing most of the successful ones, in which the feeding is done on the upper carbon rod, or in which that rod depends for its centricity upon its fitting its bushes perfectly.

The brass rods are merely wiped until bright with a piece of wash-leather and afterward polished dry. *Oil is seldom or never used to lubricate the rods*, and most arc lamps will be thrown out of action if oiled at all. It will be found to clog the action and to impede the flow of the current to the rods, and it will get carbonised by the heat and dried up by the heated air. If oil be used a good deal of trouble is in store for the attendant upon arc lamps. These remarks apply more particularly to the brass carbon rods, although they are generally true of the other working parts. The interior of the carbon rods in Brush's lamp is usually cleaned by a kind of ram-rod, using upon its end a plug of tow or cotton. If the interior has become gummy through the glycerine thickening, boiling water may be used to clear out the tube. In cleaning the brass carbon rods of lamps it is necessary to handle them so that there is no risk of bending or scratching. In the carbon rods to which these instructions apply a mixture of glycerine and water, as in the dash-pot, is used, the rod being filled when the piston is at the bottom. If the carbon rod on the positive side is correctly adjusted, it will gradually descend from top to bottom in the space of three minutes. The other rod takes a little longer. If the

ARC LAMP TRIMMING. 99

rods descend too slowly the viscosity of the glycerine is too great and more water must be used. If too fast more glycerine must be added. The rods in this and every other arc lamp must work quite smoothly and have no tendency to "stick." In frosty weather it may be advisable, to prevent freezing, to add a little pure alcohol to the glycerine mixture. As received from the makers the carbon rods are usually filled with glycerine and ready for working, save that the plug screwed into the top, to prevent escape of the glycerine, has to be removed in order to allow of the plunger-rod being raised, so that it may be hooked on to its support in the chimney cap. The cap of the reservoir is not removed save when it is required to clean the latter.

Fresh carbons are usually 12 in. in length. The upper rod is to be passed into its socket as far up as possible, and the lower carbon adjusted centrally to it, so that there is a space of at least $\frac{1}{4}$ in. between the two carbons. If the carbons are of equal diameters, as they usually are, the bottom one will have a length of 6 in. only—the top carbon burns away twice as fast as the bottom. The proportions in length must be carefully observed, for if top carbons are too short in proportion to bottom carbons, the top will be burned away too soon, and fusing of its holder will probably ensue. The carbons must always be centrally in line with each other.

Whenever it is necessary to put carbons in a lamp, or to adjust the lamp in any way, or to handle it for any purpose, the switch must first be used so as to turn the current off that lamp.

If the lamp fails to light up on turning on current, see that the carbons are touching each other—they must touch until the current begins to raise the upper carbon, on the establishment of the arc. If the automatic cut-out comes into play without apparent cause, examine the lamp to see whether the carbons are used up, or there is any obstruction to the free movement of the carbon rod.

When the arc is too long, it will look particularly blue, and have a flaming, unsteady appearance, with dullness—this shows want of proper balance in the parts, and they should be so adjusted as to give greater weight to the upper carbon, so that the raising solinoid will not have so much influence upon it.

When the arc is too short it will generally emit a hissing noise, indicating too high a temperature of the arc, and the light will be dull, chiefly through obstruction of the downward rays from the upper carbon crater. This can be remedied by so balancing the carbon rod that the solinoid will exert an increased lifting effect upon it. This adjustment, for both long and short arc, is effected in the Brush lamp by a steel-adjusting spring, which can be set at any desired position.

Arc lamps working upon alternating-current circuits must have careful adjustment for periodicity or phase of the alternating-current dynamo.

Focussing arc lamps are generally fed with an upper carbon twice as thick as the lower. This is especially the case in such a lamp as the Brockie-Pell, when made focussing. The lower carbon in this case is raised by means of cord communicators by the descent of the upper carbon. Both carbons must thus be of the same length, the difference in thickness being due to the positive (upper) carbon burning away the faster.

The diameters of ordinary carbons for general lighting, with lamps taking about 10 ampères at 45 volts, are from $\frac{3}{8}$ths to $\frac{5}{8}$ths in. These are lamps producing short arcs. For longer arc lamps, taking from 5 to 8 ampères at an E.M.F. somewhat higher, smaller carbons are used.

Arc Lamps in Series, with Incandescent Lamps in Parallel.—This system is in use in a few places of business, and even in street lighting, but it has proved itself one of the most troublesome systems yet tried. It is almost impossible to prevent the fluctuation of the arc lamps from influencing the incandescent lamps. This implies short life to the latter. Moreover, very few arc lamps can be thus inserted in a circuit. Suppose the pressure to be 100 volts, only three, at most, arc lamps could be run in such a circuit. The supposed economy of the system has led electricians to devote a good deal of time and thought to the subject. There are other fundamental defects in the system which cannot be entered upon here.

The most successful method of wedding the arc with the incandescent lamp is no doubt the ordinary parallel system, with not more than two incandescent lamps in series parallel. Each arc lamp to be in parallel singly. It appears at first the more expensive, but that is a question that can only be answered by practical trial. The young electrician, in planning for the running of arc lamps in parallel with incandescent lamps, must not forget that while his glow-lamps may at most only need an ampère each, the arc lamps will require a minimum of 5 ampères each —all of which considerations must be arranged for in the leading wires.

Arc Lighting Circuits.

Running Leads.—The simplest case in which leads can be run is that in which a dynamo machine on the ground is to be connected to a lamp elevated on a pole. It is, indeed, only a few years since—about 1880—that this was the only way to produce an arc light; each lamp had its own dynamo and pair of leads. But, as we have observed, improvements in the lamps and machines have put it in the power of the electrician to run as many as fifty arc lamps in series upon a single machine, or as many as he can find current for in parallel across the leads.

As may be expected, the insertion of a number of lamps upon the wires of a single dynamo, either in series or parallel, opens up practical questions of some little difficulty, and a great deal of trouble was encountered by the earlier experimenters in this direction. But most of the problems have been solved in the only satisfactory way—that of practical working, and the multiple series and multiple arc systems are now both pronounced successes.

The leading wires or cables are usually of copper, although iron has been used in some cases for high-tension working. They are almost invariably insulated, either practically or in name only. A properly insulated lead or cable will first be of tinned copper, covered by a sheath of pure india-rubber; then covered by a wrapping of *india-rubber prepared tape*, and, finally, a wrapping of tar-flax. The insulation may be carried much further than this, but, in either case, the wires so treated may be called insulated wires. The covered wires are merely covered, not insulated. They are usually of bare copper, with

ARC LIGHTING CIRCUITS. 103

a wrapping of cotton tape, previously prepared by passing it through some insulating liquid compound.

It will be convenient to distinguish between the two by calling the india-rubber covered wire *insulated lead*, and the cotton-covered wire *covered lead* only.

Technically, a *wire lead* is a single wire conductor. A *cable lead* is a conductor formed of several stranded wires, known respectively as wires or cables.

Insulation resistance is a point of much importance. It is usually expressed in terms of the meg-ohm (1,000,000 ohms) per mile. In tables the meg-ohm is frequently represented by the Greek letter Ω. The insulation may vary from 150 meg-ohms for ordinary insulated cable to 5000 meg-ohms for heavily-insulated and vulcanised cable. For low tension work, to be lined on insulators, the covering insulation is usually neglected, or naked wire only is used. For work of low tension, not lined on insulators, but merely laid in wooden channels, &c., the insulation giving 150 meg-ohms is generally considered safe.

For high tension work, not lined on insulators, the vulcanised rubber-covered cables are usually employed, giving 5000 meg-ohms insulation resistance per mile.

Mechanical protection is imparted to cables by hemp wrapping, wire braiding, or lead covering.

For outdoor arc lighting work cables are used in preference to wires. A wire above No. 8 gauge is stiff and difficult to handle. A stranded cable of the same capacity is much more flexible.

The legal gauge is now the recognised standard of measurement of wires in this country. It is very similar to the old Birmingham wire gauge, but is more complete, and affords a wider range.

In selecting the size of a cable consideration must be given first to the tension that is going to be maintained in the circuit. The size of the cable will depend more upon this than upon any other consideration. Let us take two opposite and extreme cases :—
(1) For a single lamp, run by a small dynamo, giving about 15 ampères at 50 volts, the usual size of cable employed in practical work is composed of seven No. 20 wires; when the distance between the machine and lamp does not exceed 500 yards, or a total length of circuit of 1000 yards. Such a cable has a resistance of 6·175 ohms per mile. But the same cable will feed several lamps, if the electrical pressure in circuit be raised in proportion to the number of lamps. (2) Forty lamps are fed by a Brush dynamo through a cable composed of seven No. 16 wires, over a total length of a mile. The volts that can be afforded as loss in the cable will always determine its size. It is a question of cost of power and cost of cable. Theoretically, the larger the cable the better.

The standards in the following table have been adopted by the India Rubber and Telegraph Works Company, whose cables are specially prepared to suit the various requirements of electric lighting work. The insulation consists of several classes, ranging within the insulation resistances per mile already mentioned. For all arc-lighting work in the neighbourhood of buildings, where the wire is apt to be handled, or to touch conductors, the cables should be insulated, whether run upon porcelain insulators or not.

Ground leakage is the most troublesome opposing factor in the work of running an insulated arc lead for high tension. It can only be obviated by good insulation, either upon the cable itself or in the form of

ARC LIGHTING CABLES. 105

GENERAL TABLE OF REFERENCE FOR ELECTRIC ARC-LIGHT CABLES.

PARTICULARS OF CONDUCTORS.

No. of wires in Strand.	Legal standard gauge of each wire.	Diameter Of each single wire.		Diameter Of the Strand.		Equivalent to solid wires. Diameter.		Area.		Weight of conductor. Per statute mile.	Per Kilometre.	Resistance at 60° Fahr. Per statute mile.	Per Kilometre.
		In.	m/m.	In.	m/m.	In.	m/m.	Sq. in.	Sq. m/m.	lbs.	Kilogrs.	Ohms.	Ohms.
7	20	·036	·914	·108	2·74	·096	2·43	·0072	4·65	147	42	6·175	3·835
7	19	·040	1·02	·120	3·04	·107	2·71	·0089	5·77	182	52	5·002	3·1079
7	18	·048	1·22	·144	3·66	·128	3·25	·0128	8·30	262	74	3·473	2·158
7	17	·056	1·42	·168	4·27	·149	3·78	·0174	11·28	356	100	2·552	1·585
7	16	·064	1·63	·192	4·88	·171	4·34	·0229	14·73	465	132	1·953	1·213
7	15	·072	1·83	·216	5·49	·192	4·87	·0289	18·66	589	166	1·543	·9589
7	14	·080	2·03	·240	6·10	·213	5·41	·0356	22·98	727	205	1·253	·7785
19	20	·036	·914	·180	4·57	·159	4·03	·0198	12·74	402	113	2·261	1·404
19	19	·040	1·02	·200	5·08	·176	4·47	·0243	15·72	496	140	1·831	1·137
19	18	·048	1·22	·240	6·10	·211	5·35	·0349	22·66	715	201	1·271	·7897
19	17	·056	1·42	·280	7·10	·247	6·27	·0479	30·91	973	274	1·079	·6704
19	16	·064	1·63	·320	8·12	·282	7·16	·0624	40·25	1,270	358	·7154	·4445
19	15	·072	1·83	·360	9·14	·317	8·05	·0789	50·96	1,608	453	·5652	·3512
19	14	·080	2·03	·400	10·1	·352	8·94	·0973	62·77	1,985	559	·4579	·2845
19	13	·092	2·34	·460	11·6	·404	10·7	·1282	83·20	2,625	740	·3462	·2151
19	12	·104	2·64	·520	13·2	·458	11·6	·1647	106·3	3,354	945	·2709	·1683
37	16	·064	1·63	·448	11·3	·394	10·0	·1219	78·6	2,482	699	·3661	·2274
37	15	·072	1·83	·504	12·8	·443	11·2	·1541	99·58	3,142	885	·2892	·1797
37	14	·080	2·03	·560	14·2	·493	12·5	·1909	122·9	3,879	1,093	·2343	·1456
37	13	·092	2·34	·644	16·3	·566	14·3	·2516	162·6	5,130	1,445	·1772	·1101
37	12	·104	2·64	·728	18·4	·640	16·2	·3217	207·7	6,555	1,847	·1386	·0861
61	13	·092	2·34	·828	21·0	·728	18·5	·4162	268·7	8,477	2,389	·1072	·0666
61	12	·104	2·64	·936	23·7	·823	20·9	·5319	343·4	10,832	3,052	·0839	·0521

porcelain cups. Next to ground leakage, the danger of *short-circuiting* is doubtless the most common. This latter is *dangerous* in two ways. Within a building *a short circuit may cause a fire,* by establishing an electric arc, or by heating a wire red hot near to woodwork. Outdoors it may burn up the armature of the dynamo and destroy the instruments. It may be remarked, in passing, that the use of special gear, flanges, &c., for the purpose of keeping a belt from slipping off a dynamo pulley is not advisable in all classes of work. *The slipping of the belt frequently saves the dynamo from destruction* when a heavy load is thrown upon it by accidental short-circuiting. A short circuit is got when the naked leads touch each other. It more generally happens through both leads getting into conductive contact with metallic substances, as girders or gas pipes. Ground leakage is generally due to wet or moisture conducting the current to earth. If the dynamo be well insulated, the tendency to ground leakage will not be so marked. The insulation of an outdoor line is generally good in dry or frosty weather, and is more likely to become faulty in wet and stormy weather.

Damages by lightning cause a great deal of trouble. During a thunderstorm outdoor leads generally give some indication of it in the dynamo room, and flashes will frequently be seen playing about the switch boards. This is usually obviated by a device called a lightning arrester or protector. It consists in many cases of two serrated plates (toothed plates are still more efficient) connected to the positive and negative leads, and so adjusted in a frame as to be face to face with a copper plate connected to a good earth plate, or other earth connection between them, very close

LIGHTNING ARRESTER.

together, without touching. The tendency of lightning is to readily discharge itself from *points and ridges* across the shortest gap *to earth*. Fusion of armature coils is the usual result of a lightning discharge passing into the line and not finding "ground." Lightning arresters should be frequently examined, because they may be disabled by coming into use unknown to the attendant, or the serrated and earth plates may even be fused together.

The well-known fact that a magnet will repel an arc from between its poles is the principle of the improved lightning protector used by the Thomson-Houston Company. The apparatus is represented in Fig. 30, and consists of a powerful electro magnet, the field of which is occupied by two metallic horns. A lightning stroke passing over the line to machine is arrested at this point and diverted to earth.

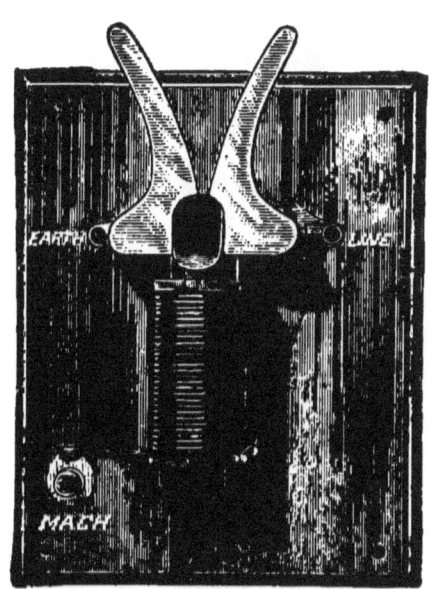

Fig. 30.—Lightning Arrester.

The use of lightning arresters is now considered essential in connection with all outdoor arc lighting lines.

Pole and Wall Insulators for the support of the leading wires are almost invariably of porcelain. For out-of-door work, especially where good insulation is desirable, as when high-pressure currents are carried, Johnston & Phillips's fluid insulator, of the cup type, has proved itself one of the most efficient yet tried. The insulator contains (Fig. 31) a small annular cup-space, containing a little resin oil, which adds enormously to the insulating power of the support. The insulators are filled with a little syphon (Fig. 32). For very high tension work the insulator is made with two spaces for the insulating oil (Fig. 33).

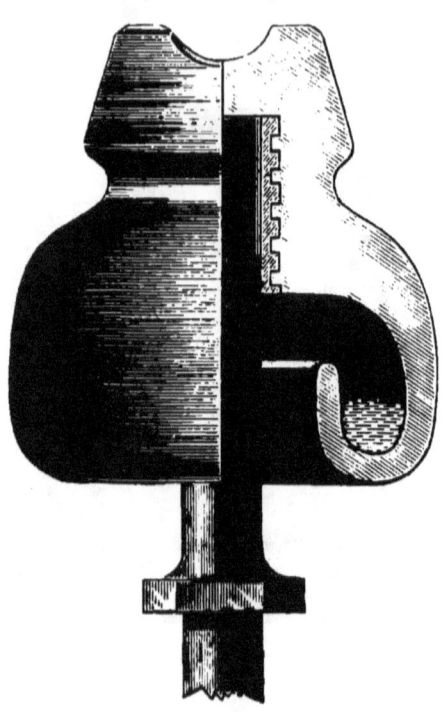

Fig. 31.—Fluid Insulator.

Heavy arc lighting leads are usually supported by independent insulators at each pole, the cable being severed, and, after being securely shackled off, reconnected electrically by a loop. In addition to these precautions, heavy leads—as main leads—are sometimes borne upon a steel rope.

This particular kind of work, known as "overhead line running," in which both earth and housetop poles are used, scarcely comes within the scope of our pages. We can only point out the more im-

INSULATION OF THE LEADS. 109

portant precautions to be taken in running leads for house lighting, and perhaps indicate the nature of the insulators and other adjuncts employed.

Naked leads must in every case be carried upon insulators. They must not be hung so near to each other that by any accident they may come into contact. A space of nine inches is the usual minimum between them. The authorities of towns will not usually allow such leads to be carried across streets. In such case an insulating covering is always required.

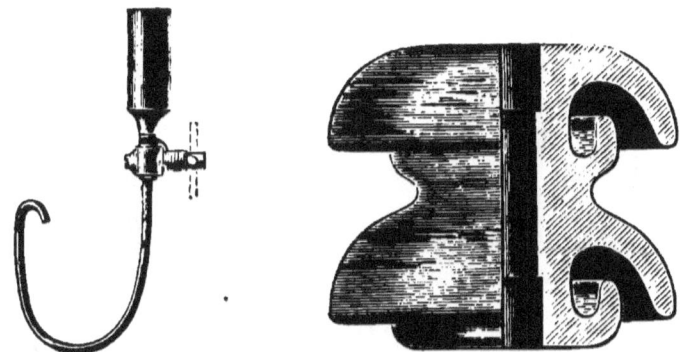

Fig. 32.—Syphon for filling Insulator.   Fig. 33.—Fluid Insulator. Double type.

Naked wires of any kind should never be carried indoors, or near to any inflammable substance.

Lightly insulated leads should be carried upon insulators. They are not quite safe when laid in wooden troughs, especially where there is any danger of the covering being abraded. If such a lead were to get a good chance of contact to earth a very little pressure or attrition would cause a short circuit.

Properly insulated leads should be carried upon insulators, if practicable, outdoors. Indoors such wires may be laid in wooden troughs, run under cleats along flooring, and so forth, but should be kept away from damp walls or iron piping. The distance between them if carried in cleats should never be less than 3 inches. Most fire-offices insist upon a greater separation. The rules recommended by the Institute of Electrical Engineers, given at p. 212, should be referred to on this point.

Heavily-insulated leads may practically be laid anywhere, so long as they are protected from mechanical injury. They are frequently laid in wet trenches (some well-insulated cables are best laid in water)

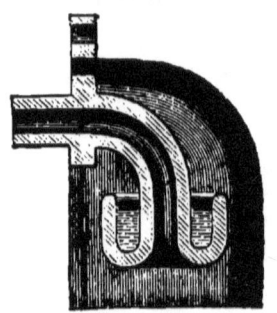
Fig. 34.
Wall Conduit Tube.
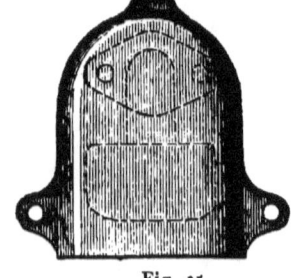
Fig. 35.

and may be carried upon damp walls. The mechanically protected leads, having a wrapping of heavy tape over all, or an armour of steel wires, or a casing of lead, are adapted for main lead work, where large currents are carried.

Danger due to Metallic Armour.—It may be pointed out that leading cables provided with metallic armour are frequently a source of danger, through the additional risk of connection between core and sheathing; as it will be discerned that such a contingency is quite equivalent to *uninsulating* the whole cable.

When a branch is taken off through a wall into a house the aperture must be made in the wall *above* the terminal insulator. A porcelain wall conduit-tube,

INSULATION OF THE LEADS.

Figs. 34, 35, which represent the conduit in section and front elevation, and furnished with the fluid insulator, must be used in passing the lead through the wall. If there are two leads the pair must not pass through one tube, unless it be divided by an insulating partition. The wire is carried upwards to prevent rain and wet from following the surface into the wall.

The apertures in partitions and interior walls must in every case be lined with either porcelain tubing or vulcanised rubber.

In high-tension arc lighting within mills, stores, and so forth, the lead should always be exposed to view, and run upon insulators several—many offices insist upon 12—inches apart. If a wire must be concealed it must be heavily insulated at that point.

In low-tension working, as when the arc lamps are run in parallel with a pressure not exceeding 50 volts, the leads, if well insulated, may be run upon woodwork and fastened by wood cleats or leather loops.

The making of *joints* and *splices* will be found more particularly described in Chapter V., p. 190. Joints must be good *mechanically*, so that their breaking strain is greater than that of the wire.* They must be made good *electrically* by soldering. Every joint must be heavily insulated after it is complete by a wrapping of prepared tape.

Planning a System of Mains and Feeders.—An admirable system of preliminary planning in getting out a network of main leads, with the necessary feeders for keeping the potential equal at all points, comes from Berlin. The German Edison Company take a large frame or table, and make a clear plan

* The ordinary telegraphic mechanical tension-resisting joints are fully described at p. 302 of "Electric Light," 4th ed.

thereon of the streets, buildings, &c., to be covered by the system. The location of the central station is then marked, and two wires from a small battery run to it to represent the electric supply. From this point the main leads are all laid down in miniature, along the plan of the streets, and each group of lamps is represented by a wire resistance. Current is kept upon the system, and the drop of potential at each point carefully noted by galvanometer. Feeders are then run to equalise the potential, and by means of careful measurement every detail of the system can thus be ascertained. The model network is kept at hand, and as any alterations are required in the real net-work they are first made upon the model. This is of course much better than any possible paper-and-ink system of planning.

English engineers generally get out their plans upon paper, marking off the lengths of cable and location of lamps, and calculating the carrying capacity of the main leads upon the basis of 1000 ampères per square inch sectional area of leading wire. These systems have, however, more particular reference to incandescent lighting, and the reader is referred to Chapter V. for rules and tables from which can be ascertained the best size of conductors to employ for any particular installation, or number of lamps, at various distances, with various losses upon resistance.

Transformers or Converters.

The fundamental conception of the modern transformers is represented diagramatically in Fig. 36, where D is a dynamo, in whose circuit the primary coil P is placed. The electromotive force in this circuit is supposed to be high, say 1,000 volts. S represents a secondary coil wound in contiguity to P, but insulated therefrom and in no way communicating except by induction. In the circuit of S are the lamps, arranged in the usual parallel way. The dia-

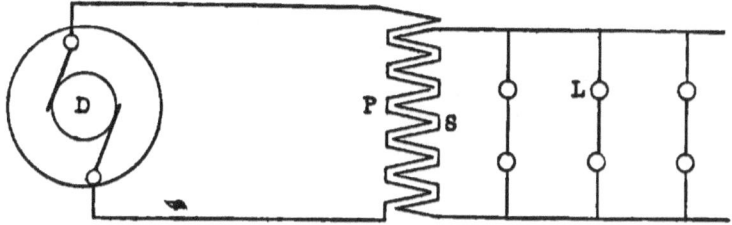

Fig. 36.—Diagram of Transformer.

gram shows two lamps in series. P S is then the transformer, the function of which is to convert the 1000 volt E M F to, say, 100 volts. In a practical transformer a core system of soft iron would occupy

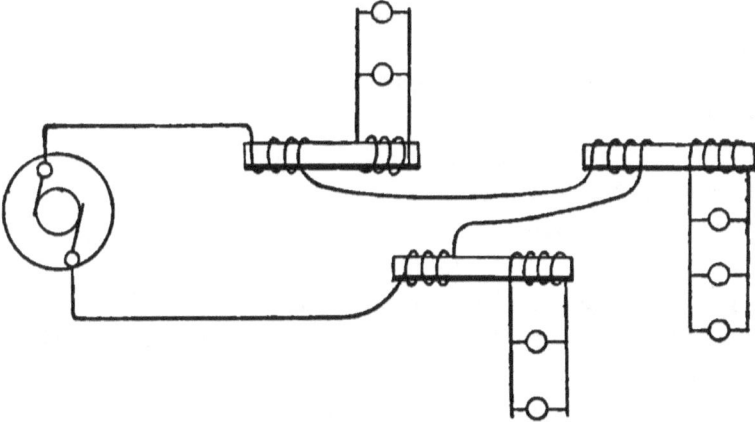

Fig. 37.—Diagram of Transformers in Series.

the centre of the two coils. In Fig. 37, where the transformer is symbolised by an iron bar with primary and secondary coils, the conception of running converters in series is represented, each transformer feeding its own lamps.

I

The Working of Transformers.—Transformers are run upon alternating current circuits. Their employment is essential when a high electromotive force is put upon the main leads. The common voltage of the alternating currents is seldom less than 1000. For use within buildings, or for arc lights, this pressure is reduced by a transformer to 50 or 100 volts, as the case may require. In practice engineers always employ as *small* a transformer as possible. The use of a transformer that is too large causes great waste of power. For this reason transformers are now made in many sizes, to feed from 10 to 100 lamps.

The construction of the transformer or converter cannot be treated in these pages; it is amply dealt with in descriptive works on electric lighting. Fig. 38, however, represents the external appearance of a Thomson-Houston transformer, where the heavily-insulated cables represent the high tension primary coil, and the lightly-insulated wires the 100 volt house circuit.

The nature of the *induction coil* is too well known to call for any description here. A transformer is a slightly improved induction coil, built to take a large current, and in a compact form. The make-and-break contact and the condenser of the inductorium are absent from the transformer, since alternating currents are used in the primary,

Arc and incandescent lamps are frequently run off the same transformer. In some cases two arc lamps are inserted in series across the mains. This system is bad and troublesome. The lamps will probably be unsteady, and if one goes out the other must follow. One lamp across the mains, steadied by an impedance coil, will be found much more satisfactory. Before proceeding, let us examine the nature of

Impedance and Choking Coils.—Choking coils are used in connection with alternating currents, and consist of inductive resistances in the shape of coils of closely-wound copper-wire, the effect being much

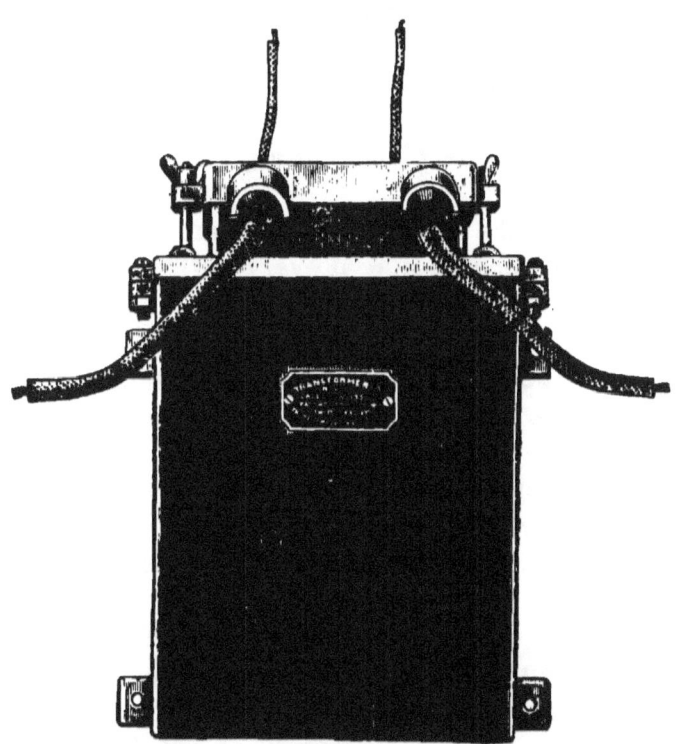

Fig. 38.—Thomson-Houston Transformer.

increased by the presence of iron within the coil, and greatest when that iron forms a short, closed magnetic circuit. The object of such a coil is to obstruct or choke any initial E M F without the loss of power which an ordinary resistance would entail; and in connection with alternating arc lamps choking coils are employed exactly as resistances are used when running lamps in parallel; with this advantage, that, if well constructed, a much smaller loss of energy takes place; moreover, should the lamp get short circuited, the consequent increase of current is much less than if the ordinary resistance is employed, as the self induction of the coil increases with the current, and so retards or obstructs the E M F, and therefore the current does not rise so high.

Such coils are usually too bulky to be placed in the lamp itself, and are placed in any convenient part of the circuit.

The higher the periodicity the smaller the coil may be made, so much so that at every high periodicity the coils of the lamp itself may form an inductive resistance sufficient to enable the lamps to be worked in parallel without any other outside resistance.

Impedance coils are generally considered identical with choking coils, simply a difference of name, but the term would be more suitable for the inductive resistance sometimes used in running arc lamps in parallel on continuous current circuits. The chief advantage of such coils is to prevent the great development of current in an arc circuit at the moment of completing it, thus preventing the violent action of the regulating mechanism of the lamp, and which causes much of the pumping of arc lamps when worked in parallel.

The resistance in ohms of a fairly constructed choking coil would be about one-twentieth of an ordinary resistance having the same reducing result of current, with a given E M F; but a much wider difference can be obtained if the magnetic circuit is made very low, or the periodicity very high.

If the 100-volt pressure, usual in this country, be employed, a "choking coil" will have to be inserted in that bridging wire to feed the lamp at the proper pressure. It is a very general practice on constant current parallel circuits to use a "choking coil," even with single lamps, across the leads. The resistance due to the choking or impedance coil generally causes a damming back of from 10 to 15 volts in the pressure. Although better results are thus obtained than by merely inserting the lamp alone, it leads to considerable waste where a number of arcs are burned upon a circuit. One great advantage of the alternating current is the avoidance of much of this loss. When arcs are used the lamps can in most cases, on 50-volt circuits, be inserted direct along with a small impedance coil. On a 100-volt circuit a loss by the reason of "choking coil" is inevitable, but an impedance coil fortunately uses much less energy than a resistance.

The following are the resistances of the coils used with the Brockie-Pell arc lamps for parallel working: In constant circuits $\frac{1}{2}$, 1, 1$\frac{1}{2}$, and 2 ohms, mounted in a frame or cylindrical case, the various resistances being adapted to various pressures. Impedance coils for alternating current working are adjusted on the spot until the necessary steadying effect is obtained. They are usually fitted in a separate case near to the lamp. We understand that Mr. Brockie is sanguine of being able to so arrange the circuits of his lamp as to obviate the necessity for the employment of impedance coils.

While 50 volts is usually called for by an arc lamp upon a constant current circuit, from 35 to 45 volts is found to give an equally satisfactory arc upon alternating current circuits.

"*Singing*" *of the Alternating Arc.*—Alternating current arc lamps usually emit a humming noise, rendering them unfit for indoor lighting, unless in a very wide and high space.

It may be worth while to note that the alternating current arc, since it does not produce a crater in the top carbon, casting the light powerfully downwards, always needs to be covered by a superior reflector.

Location of Transformer.—Where it is at all practicable the transformer should be kept outside a building, but must be protected from wet or damp. It should be fixed in a fire-proof place. It may be explained to those not familiar with transformers that they are usually rather dangerous pieces of apparatus, from two causes. They are always used where very high-tension currents are employed, and form a kind of terminus to those currents from the generating station. Hence a transformer should not be handled by an inexperienced person. In addition to this transformers always become warm, and sometimes hot, by reason of the continuous reversals of current through them, or to what theorists have called "magnetic hysteresis." This heat should not rise above 150° F., but it sometimes, owing to faulty construction or other defect, rises much higher, and the transformer may possibly get hot enough to char dry wood. In addition to these undesirable attributes a transformer frequently emits a singing noise, or is seldom or never quite silent when the current is on.

Transformers are usually placed in cellars, through the want of any accommodation elsewhere. In such cases they should be mounted upon iron brackets, at a distance of not less than 12 inches from any wall, and as far as possible from any inflammable substance. The terminals and high-tension wires should be guarded from any accidental personal contact. In the case of the transformer being supplied from street mains, a double cut-out switch should always be located outside the building as well as within, for use in case of fire or other contingency.

A new transformer that will not act will probably be found to be damp. The most ready means of drying is to pass a pretty strong current through it for some hours. It is usual to connect the framework and all the iron portions to earth. This precaution will earth any leakage from coils to core, and render the transformer comparatively safe when touched by the hand. In some instances iron cases, providing ample room, are furnished for transformers, especially for outdoor work.

In America it is common to see the transformer belonging to a building fixed upon the street line wire pole opposite.

Transformers in Parallel and Series.—For parallel working of the transformers *pressure* must be kept constant in the primary wire. It is very unusual to put more than one transformer across the mains. For series working the *current* must be maintained uniform.

In practice it is found that in series working it is not advisable to leave the secondary circuit open. It is therefore a general practice to short-circuit it when idle. Series working is rather rare. On the other hand the secondary should never be short-circuited in parallel working. It is generally open-circuited by the simple operation of switching off the lamps.

The usual loss of pressure in a house transformer is

very generally as high as 2½ per cent. of the pressure supplied to the house circuits. It depends, however, upon the efficiency of the transformer. Some transformers exhibit an efficiency as high as 95 per cent., others as low as 50 per cent.

Main Safety Fuses for Transformers.—A double fuse capable of carrying the full current supplying the transformer is inserted in the branch leading from the mains to the house. In addition to this it is usual to put the secondary circuit to earth. Any accidental contact, then, between primary and secondary would not convey the high pressure current into the house wires. The short circuit so formed would fuse the cutout and stop the supply. This is considered a perfectly *safe* plan.

Necessity of opening Primary Circuit when Lamps are idle.—Recent experience has shown that a good deal of loss is incurred by the generating station on account of transformers left in parallel across the mains during the hours of idleness. It would be similar to a leakage of gas from a gasmeter if the service cock were not turned off during the day. Owing to this there is a very general movement towards getting into working form a system of opening the main circuit to a house when the lamps are switched off. It may, of course, be effected by an automatic switch, but it is far more likely that a hand switch will prove the more reliable. The electricity user will be expected to do this for himself, and will no doubt attend to it, *if the supply meter be upon the primary circuit of the transformer.*

Meters for Recording Electricity Supply.—The use of meters is now general. The construction of each kind may be studied in a work descriptive of such apparatus, and does not concern us here. The electricity meter is usually fixed near to the transformer, when a transformer is used. It is merely put in circuit; or—on the three-wire system, when Edison's three-wire meter is used—in circuit of the three wires. The use and "reading" of each meter calls for a special description, which generally accompanies the instrument, and cannot be treated in these pages. "Station meters" are recorders of a larger kind, for indicating the total current passing out of a supply station; or each main may have its own indicator. The station meter is a check upon a large number of house recorders. In connecting meters the leading wires should not be carried around them (so as to encircle the meter), or in any way so disposed that any inductive effect may disturb the working of the meter.

House Main Switch-board.—In addition to the various hand switches, fitted to, or in connection with, single lamps, or groups of lamps, it is convenient to provide main switches for controlling each circuit. The "main leads" terminate here—that is the leads either from a transformer or from the street mains direct. The several circuits are so arranged that all their positive and negative wires may be put in contact with the positive and negative poles of the leads, so as to distribute the current between them. Any circuit not in use may then be switched off. Double-pole switches should, if possible, be used. By means of this method of working any danger from idle circuits is obviated. Such a switch-board in a theatre is in constant use during the performance. In small installations, where the number of lamps does not exceed 50, such a switch-board will not be required unless the 50 lamps be put upon more than one circuit, an unusual and unnecessary practice, considered electrically. The conveniences of such

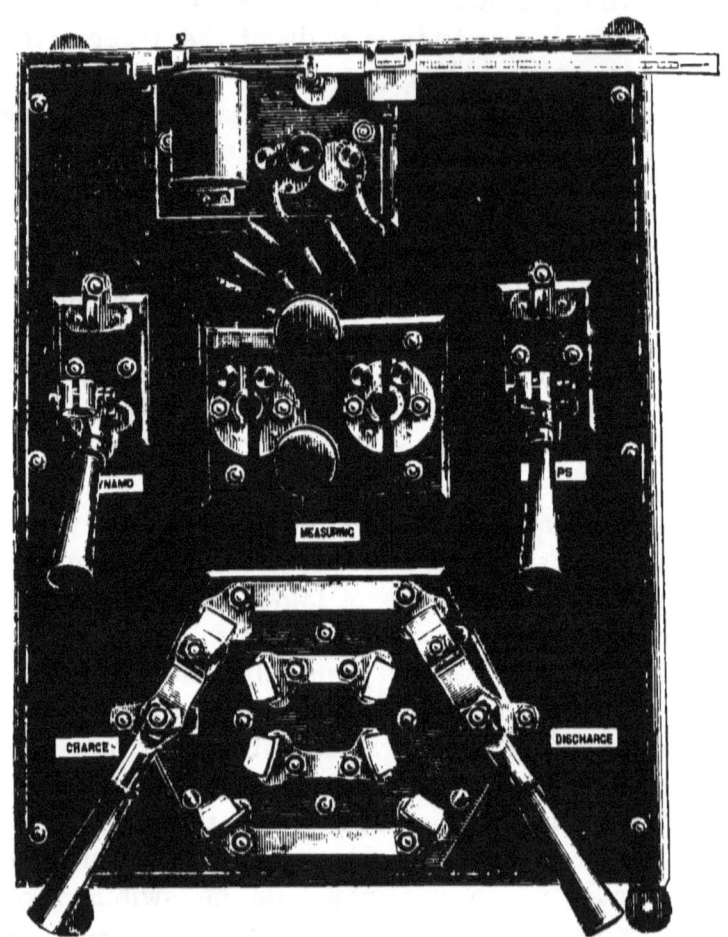

Fig. 39.—Dynamo Room Switch-board for Accumulators.

a system of switching are in many cases very considerable.

Dynamo-Room Switchboard.—Most of the private installations fitted up in banks and offices—as at the

Bank of England—are provided with accumulators. For the regulating of such plants Messrs. Drake and Gorham have designed the form of switchboard represented in Fig. 39, which is fitted with a "Steelyard" ampère meter, and Ring contact switches (see p. 151) for controlling the number of cells in action. The diagram is self-explanatory.

CHAPTER V.

WIRING FOR INCANDESCENT LAMPS.

WHEN a building is to be wired for incandescent lighting several leading questions arise. (1) The number of lights required; (2) the power of the lamps, 8, 10, 16, or 20 candle power, as the case may be; (3) the number of circuits required—when there is not more than 50 lamps one circuit may be made to serve, unless more than one is desired by the owner of the building for convenience in switching off; (4) the volts (*e.g.*, electrical pressure) required to run the lamps; (5) the ampères (current) to be consumed by the lamps; (6) the lengths of the wires, their sizes, resistances, &c.; (7) the volts to be "lost in the wires" (fall of potential due to the resistance of the wires).

The answers to these questions will not be invariably the same. They will depend upon some of the following conditions :—

Whether the lights are to be run off a private dynamo, selected for the purpose, with or without an accumulator.

Whether they are to be run off street mains kept charged by a public supply company.

It may also modify the case whether continuous or alternating currents are employed.

The use of arc lamps, as in a shop, where the inte-

PARALLEL SYSTEM OF WIRING.   125

rior is lighted by incandescence and the outside by a few arc lamps, may also have to be provided for.

Private house wiring calls for the minimum of preliminary calculation, shops and offices for greater care, whilst theatre lighting is perhaps the most difficult of any to scheme.

The practical carrying out of the wiring itself is only to be effected upon one plan, as regards quality of the work, that is, the best possible method according to present knowledge.

The number of lights will be approximately as the number of gas jets of 16 candle gas that would be required, and should in every case be ample for the purpose.

The System of Wiring.

Parallel, or "*Multiple Arc.*"—A great deal of the best lighting in this country has been carried out upon the parallel system. It is very convenient. It is easily understood, and it is a very safe system.

The simplest idea of the parallel system that can be given is represented in Fig. 40, where D is a dynamo,

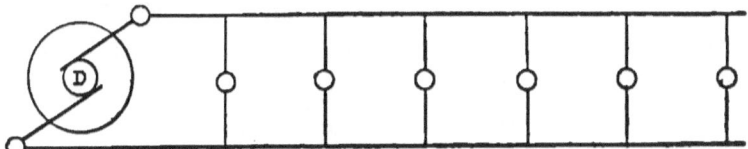

Fig. 40.—Diagram of the Parallel System.

or other electric source (it may be the mains from a lighting station or an accumulator). Two wires are led from the + and — (positive and negative) poles. They are kept apart, and are insulated. The lamps to be fed are then connected to the mains with insulated wires, as shown. The current divides itself between the lamps, lighting each and every lamp with equal effect. But if too many lamps are put across the mains a fault will soon be apparent. Those near to the dynamo will get too much electricity, and those at the far end too little. The lamps nearest to the dynamo will thus be abnormally bright, and will, soon be worn out, while the distant lamps will have the appearance described as similar to a "red-hot hair-pin," giving insufficient light.

This falling-off in the light given by the distant lamps will only be felt if the two main wires are very long. In practice the two main wires would not be connected to the dynamo as shown unless a few lamps only were to be run from them. If there were *fifty* lamps upon the wires, arranged as shown in the diagram, the distant lamps would certainly not be so bright as the nearer lamps. If there were 100 lamps the far lamps would be very indifferently lighted, and in this way the force would be drawn away from the mains until a point would be reached at which little or no electricity would be present to cross from one main to the other. This defect would be obviated by connecting the dynamo to the central portion of the wires, not to their ends. If 20 lamps could be run off a pair of mains, as represented in the diagram, 40 could be lighted equally well by simply disconnecting the dynamo from the extremities, and connecting it to any point near the middle of the line of lamps.

If there were still a larger number of lamps, and the distant lights showed insufficient current, branch leading wires from the mains or feeders would be led, as represented in Fig. 41, which is the distributing

FALL OF POTENTIAL. 127

system first introduced by Mr. Edison. In this way every lamp will burn equally bright. If even half the lamps be turned out the other half will not perceptibly brighten up. The turning off of any lamp will not affect the lights in the next room. The question of street mains and feeders does not concern us here, but it may be as well to bear in mind that the distribution of electricity within a building is dependent upon precisely the same conditions. Although *mains*

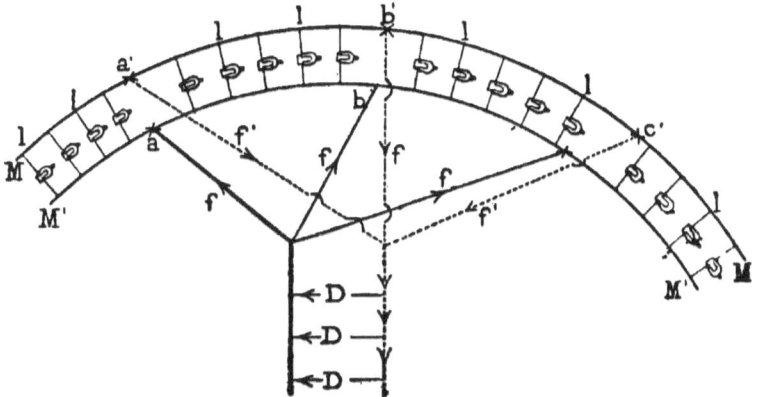

Fig. 41.—Diagram explaining the Use of Feeders.

and *feeders*, as such, may not have to be arranged for, yet the electrical "pressure"* must be effectually equalised.

Fall of Potential or "Pressure."—Let us examine this a little further. It is a well-known fact that if the two mains led from the poles of a dynamo, as in the diagram (Fig. 40), be very long, the electrical pressure between them will fall off in proportion to

* The word pressure is not strictly correct as applied to electricity, but so many practical electricians (many of them well-known authorities) use it and it is so easily grasped, in its analogy, that we make no scruple to employ it in these notes, addressed as they are to practical wiresmen.

their length, if the lamps be equally distributed. If the lamps be grouped at certain points it will be more apparent at these points.

Fig. 42 and Fig. 43 are two familiar cases. The first shows a gas-pipe carrying a moderate pressure of gas, M M'. The first burner will receive the most gas, and the last the least. No two burners will be exactly alike. The second diagram represents a

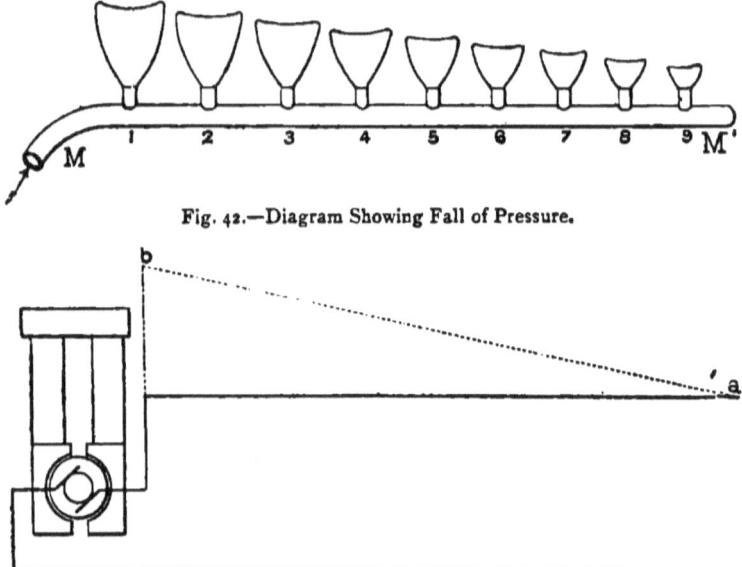

Fig. 42.—Diagram Showing Fall of Pressure.

Fig. 43.—Diagram Relating to Fall of Potential.

dynamo, the positive (or "feeding") pole of which is connected to a long main. If the electrical pressure is represented by the line $a\ b$, it may be conceived to fall off gradually as the distance is increased, as shown by the sloping line. This fact applies to all circuits. In the diagram the return half of the current is represented as flowing back by the earth, as in telegraphy, by the means of earth contact plates. The earth is

FALL OF ELECTRO-MOTIVE POWER. 129

generally regarded as presenting no resistance to the current similar to that offered by a wire, but it is seldom or never used in electric lighting, on account of other considerations explained further on.

This fall or drop of potential or "pressure" is due, in the first place, to the resistance of the circuit. The greater this resistance the greater the fall of pressure. Again, with a given resistance of wires the pressure falls off in proportion as we increase the current. This is the varying pressure that must be met, as explained in Chapter I., p. 27, by increasing the pressure at the dynamo. A certain fall is due also to leakage.

In plain words, the resistance causes a loss of working power, and the switching on of more lamps also causes a fall of pressure. The loss due to resistance is great or small according as the wires are small or large, or long or short. A wire of a certain gauge, 100 feet long, will incur a loss of pressure just twice that due to a wire 50 feet long. A wire of a given *sectional area* will incur twice the loss due to a wire of *double the sectional area*. In circuits where the wires are of even thickness, and the lamps equally distributed, the fall of potential will be even. In others, where the sizes of the wires vary, or the lamps are unevenly distributed, the fall of potential will be greatest where the wire is thinnest and where there are most lamps.*

A certain loss in leading the current to the lamps is unavoidable. Practical men know fairly well how much loss they can afford. The thicker and shorter the

---
* To enable the reader to fully grasp this part of the subject, careful study of the laws of the circuit may be said to be essential. A simple enunciation of Ohm's law, bearing upon this question, is given under "Estimation of the Electrical Power," further on (p. 184).

K

wire the smaller the loss. Wires are selected to keep this loss down to *two and a half* per cent. of the total. A practical wiresman will say that he loses 5 volts from dynamo to lamps—that is, $2\frac{1}{2}$ volts in the main leading to the building and $2\frac{1}{2}$ volts in the lamp circuits. To run 100 volt lamps his dynamo must yield at least 105 volts. $1\frac{1}{4}$ per cent. of this will probably be due to the resistance of the "*leading*" wires, and $1\frac{1}{4}$ per cent. to that of the "*return*" wires.

The leading wire, or briefly "lead," is usually that representing the positive terminal of the dynamo, sometimes called the positive or feeding wire. The return wire is that leading to the negative terminal of the dynamo, called briefly the "return" or negative wire.

If the above 105 volts be absorbed in lighting the circuits of lamps, 100 volts will be lost in the lamps themselves.

The proportional fall of potential (the volts lost) is less and less as the pressure in the circuit is increased; but the highest safe pressure for use in houses is not above 200 volts, and lamps are not constructed for ordinary candle powers taking so much pressure as this. Thus, while a No. 16 wire can be made to safely and economically carry *six* lamps, requiring a pressure of 100 volts, the same wire would not be used for more than four lamps requiring 50 volts only. These facts will, however, be better understood by rules for calculations given further on.

The Series Multiple Method.—The great advantages of the parallel system we have just spoken of are that each lamp is independent, and that a safe pressure can be maintained on the wires. Thus the breakage of any lamp, or the switching of it off has no effect upon adjacent lamps. But the economy in conductors and energy to be derived from the use of higher potentials has brought into use a system of *connecting more than one lamp* on a wire between the mains (Fig. 44). It is usual to connect two lamps in this way, and sometimes three or more. The diagram represents a pair of mains with a pressure of 200 volts, and shows that four 50-volt lamps may be arranged upon them in series, or two 100-volt lamps. The great disadvantage of this plan is the certainty that if one lamp should break or be switched off, the other must also cease to burn. It is true that

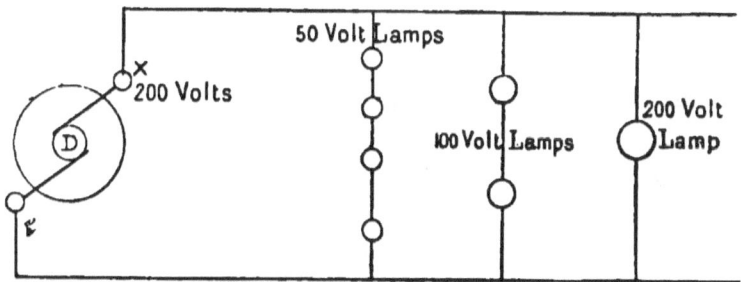

Fig. 44.—Diagram of Series Multiple Circuit.

this fault of the system has been combatted by an electro-magnetic switch, which keeps the circuit open, or provides a by-path for the current in the event of a stoppage of one of the lamps. But the use of such apparatus complicates the case very greatly, and introduces other troubles, the worst of which is the risk of fire. Besides these disadvantages, where this system is in use there is no saving of energy when half the lamps are switched off, because the magnetic switch has to insert a resistance into the circuit as great as that of the supposed broken lamp.

132   WIRING FOR INCANDESCENT LAMPS.

The multiple series system is used chiefly for large groups of lights, as in shops or theatres, where all the lights are required together. A useful and convenient combination of the parallel and series parallel may be made by bridging the main leads with two lamps in series as in the diagram. Thus, if the ordinary parallel lamps in the building are 100-volt lamps, the light at any point may be split into two separate portions by using two 50-volt lamps in series. As a matter of course, if one of the lamps should fail, the other upon that bridging wire will cease to burn. One switch serves for both lamps.

The Three-Wire System.—This is a system that has been brought into use more especially for main distribution work. It is chiefly employed for street mains. It may occasionally be utilised in large buildings. But for general wiring the three-wire system is not necessary.

Its main object is to effect a saving of copper conductor, and as this saving is very large the three wire system is coming into favour. The practical wiresman may have something to do with the system, if not in wiring a building, yet in making connections to it, and it may be well to briefly examine it. In the three-wire system the volts used are twice that employed on any parallel two-wire system, and the current (ampères) is only one half that used in common parallel.

In the working of the three-wire system two dynamos are used, connected to three conductors, as in Fig. 45. The dynamos are joined in series, and the central wire would therefore appear to be an idle wire. But when lamps are connected in the system they should bridge from negative wire to centre and from positive wire to centre, alternately, as represented. The inventor in this country (Dr. J. Hopkinson, F.R.S.), however, intends his three-wire plan to apply to *alternate houses* in a street. The central wire may be of much smaller section than either of the other two, as it has only to carry the *difference* of current between the two divisions of consumers,

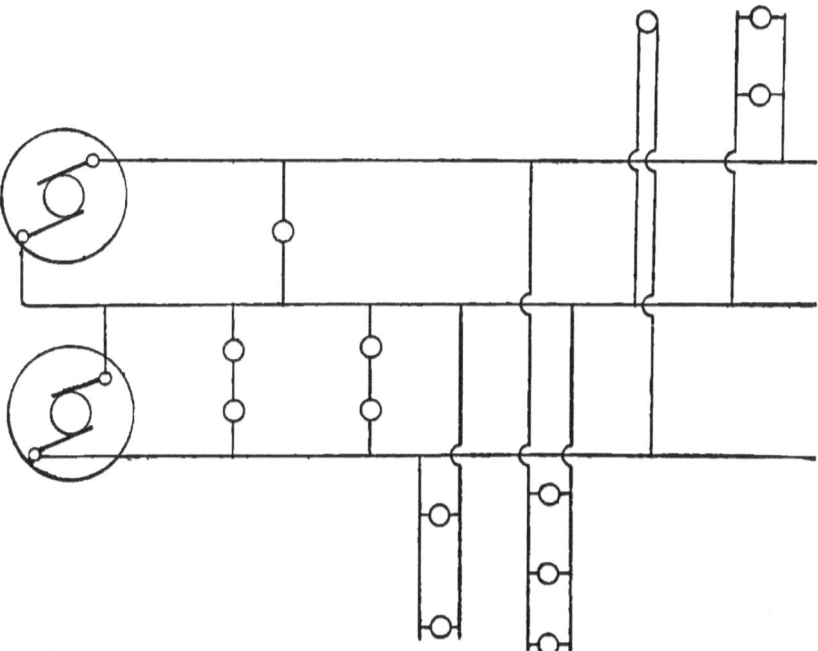

Fig. 45.—Diagram of the Three-wire System.

and is frequently an earthed conductor—that is, uninsulated; but this latter plan is not resorted to unless the two divisions of consumers require an approximately equal supply.

While the three-wire mains carry a current at, say, 200 volts, only 100 volts enter the houses when connected as shown. Therein lies the advantage of the system in respect to ordinary house wiring.

Series System.—This system of wiring is seldom used. Indeed, it is quite impracticable when carried further than a few lamps. It will be observed from the diagram, Fig. 46, that the lamps are merely connected one after the other, the whole of the current passing through every lamp. A line of 50 lamps thus con-

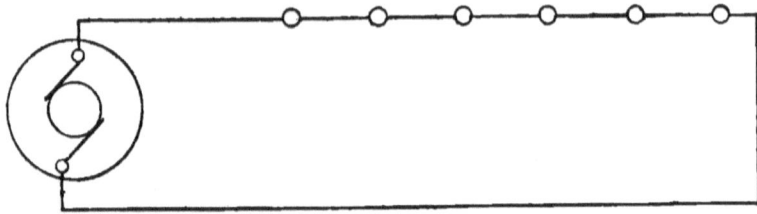

Fig. 46.—Diagram of the Series Method.

nected, if each lamp used 100 volts, would call for a pressure of 5000 volts. The system has merely been experimental, and it presents the very great disadvantage that a failure of any lamp breaks the circuit.

Multiple Series.—This is a more practicable development of the same idea as that just described. It is frequently employed when arc lamps are to be run

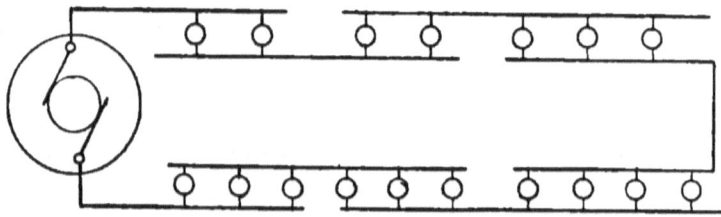

Fig. 47.—Diagram of the Multiple Series Method.

in connection with incandescent lamps. Fig. 47 represents the arrangement diagrammatically. It will be noticed that if one or two lamps should break in this system, the circuit would not be interrupted. The other lamps in that group would become much brighter, consequent upon their having to pass on the same current as was before carried by a greater number. All the lamps in one group must be 100-volt or 50-volt lamps, as the case may be. If the lamps are all of equal voltage, the number in each group should be the same.

In this system the *current* must be kept constant, and a series-wound dynamo is generally used. The current will vary as the number of lamps, and the pressure as the number of groups upon the circuit. Hence, if the number of lamps varies, the current must be increased or diminished to correspond. If the number of groups varies, the pressure must be adjusted in proportion. In the diagram the circles represent incandescent lamps; but if it be required to burn an arc lamp, it can be effected by inserting it in place of one of the incandescent lamps. Thus the arc and incandescent lamps are run upon this system in parallel. The arc lamp will of course take a great deal more current than the incandescent lamp, and may replace several of these in a group, or even a whole group. Arc lamps to run on series parallel wires must be furnished with automatic cut-out or by-pass, so that the current is not interrupted on a failure of the arc. Only differentially (shunt) governed lamps should be used. The running of the system calls for considerable care on the part of designer and attendant, and an impedance coil is required.

Working Off Transformers.—The diagrams (Figs. 48 and 49) represent the usual arrangement of running parallel circuits off convertors, alternating currents being employed. The diagrams are self-explanatory.

136   WIRING FOR INCANDESCENT LAMPS.

Lamps in Parallel.—In Figs. 50 and 51 (p. 138), D represents the continuous current dynamo, M S, main switch and fuses, S, switches and branch fuses, L, lamps run in groups upon branches from the mains.

Other Systems.—Several others have been tried. Most of them have never merged from the experimental stage. Broadly speaking, there is one system used for the running of incandescent lamps—parallel

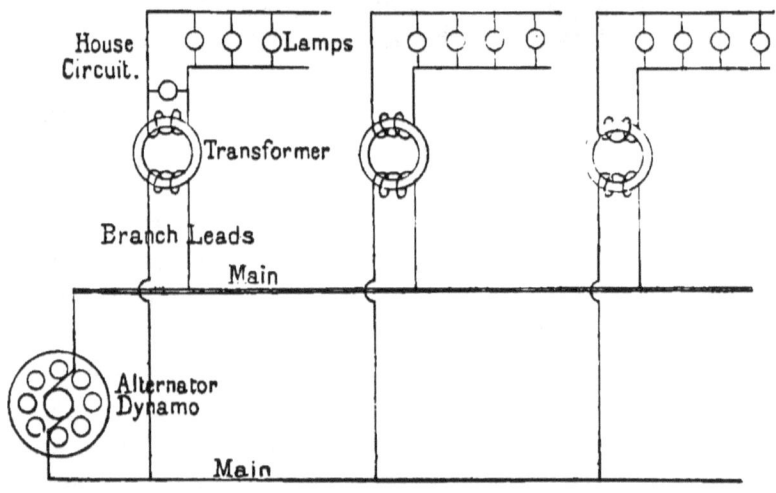

Fig. 48 —Diagram of Transformers in Parallel.

or multiple arc—and although for special purposes one or other of the different systems spoken of in these pages are occasionally used, the parallel plan seems likely to hold its position as the first, simplest, and best.

Selection of a System.

Alternating v. *Continuous Currents.*—Considerable experience in this country of both systems does not appear to show that one system has any advantages over the other. There is a certain wastage of the carbon filament of lamps in both cases. The wastage

SELECTION OF A SYSTEM. 137

is impartial with the alternating currents—it appears to occur equally over the length of the filament. With the continuous current it is partial to the positive connection of the lamp, and this end of the filament waxes

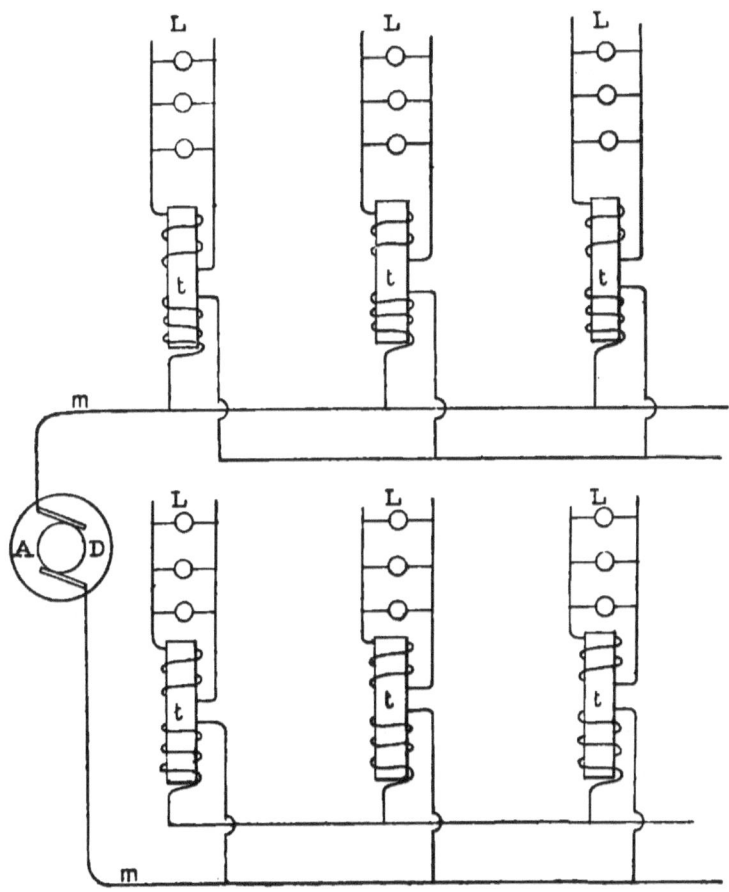

Fig. 49.—Diagram of Transformers in Parallel.

thinner than the other—the final rupture usually occurring at the end joined to the positive lead.

When lamps are run off street mains, and a *transformer* is used, the alternating currents are always

138   WIRING FOR INCANDESCENT LAMPS.

employed. When the street mains convey continuous

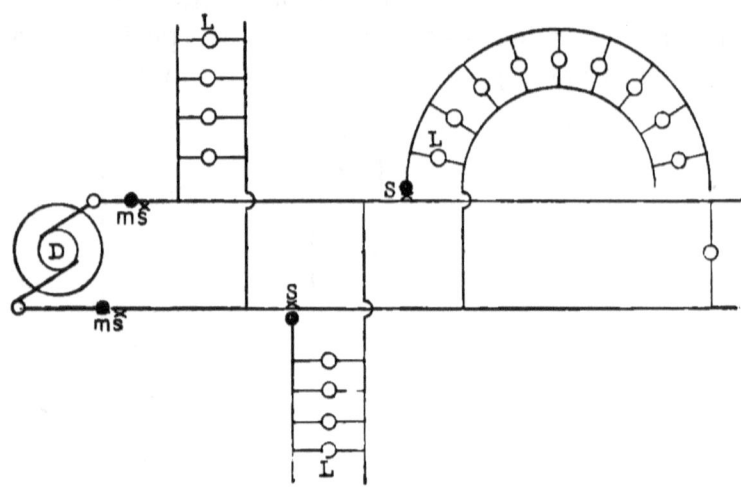

Fig. 50.—Diagram of the Parallel System of Wiring.

currents, the same currents are always used in the houses. When lamps are run off an accumulator

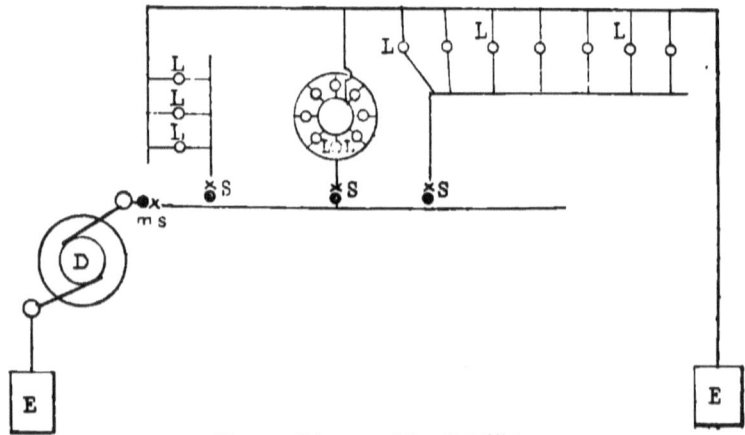

Fig. 51.—Diagram of Parallel Wiring.

continuous currents are employed. When they are run off a dynamo direct the currents may be either continuous or alternating, according to the nature of the machine. The balance of opinion appears to favour alternating currents for incandescent lamps.

Parallel or Parallel-Series.—If the voltage of the street main be 100 it is the general custom to use the ordinary parallel system with one lamp across the main wires, as already explained. If the current entering the house have a potential of 200 volts, it is common to put two such 100-volt lamps across the wires, in series, as shown in the diagram given on p. 131. It is not advisable or usual to put more than two lamps in series in houses. If it be desired, 50-volt lamps can be run in pairs in series across wires at 100 volts.

When a Dynamo is Used.—In isolated plants it may be said that it is scarcely sufficient to rely entirely upon the dynamo. It is much more satisfactory to couple with it an accumulator. If 50-volt lamps be employed, 26 cells of accumulators will be required. The *number of lamps* such a battery will run will depend upon the *size* of the cells. Taking each lamp roughly at 1 ampère, it will depend upon the current in ampères evolved by the cell at an economical rate of discharge. The rate of discharge being estimated at about 4 ampères per positive plate of the large "L" type of E.P.S. cell (p. 46), the total discharge is equal to 50 ampère hours per positive plate, so that each positive plate would discharge about 4 ampères for 12 hours. There being 7 plates the total discharge will be from 25 to 30 ampères. Hence, from 25 to 30 lamps, taking approximately an ampère each, can be run from the battery we have supposed. In estimating the number of cells required for lamps of odd voltages, divide the number of volts by 2, and add two cells as "reserve" (p. 51).*

The continuous-current dynamo must yield a current of sufficient strength to charge the accumulator. Further instructions will be found at p. 45.

An alternating current dynamo in an isolated plant will work direct on to the lamp circuits. There is this little advantage in the alternating dynamo, it is less liable to faults of conduction. Having no commutator or commutator brushes, breakdowns are much less frequent.

In making suggestions for selecting a system of working it is impossible to enumerate in a book all the possible variations from the fundamental rules already laid down. The reader must carefully consider his ground. In wiring houses for lighting from street mains the system and voltage are already there. He has only to lay his circuits, and choose his lamps to suit. In selecting an isolated plant, or a ship plant, he will be led by the requirements of each case; the balance of opinion is in favour of having an accumulator in reserve, especially in house lighting. Theatres are lighted both without and with accumulators. If they are not used, ample spare machinery should be provided in case of emergency, and in either case the circuits are frequently laid in duplicate, one set being kept in reserve; this latter precaution applies especially to the auditorium.

Planning of Circuits.

Broadly speaking, an installation of 100 lamps should be divided into at least two circuits. In many

---

* To understand the grouping of the accumulator cells a knowledge of the laws of the voltaic circuit is essential, for which see a good text-book. In simple cases the figures given by the makers of the cells are ample.

DISTRIBUTING BOX SYSTEM.

cases 50 lamps is too large a number to place upon a single pair of wires. The result will be more satisfactory, in respect of the electrical distribution, if three or more circuits are planned for.

Low voltage work—50 volts and under—is unusual in this country; 100-volt lamps are the rule. When the wire section is not restricted high voltage circuits will carry a greater number of lamps than circuits with only 50 volts.

The Distributing Box System.—According to this plan of arranging the wires all the switches (save in-

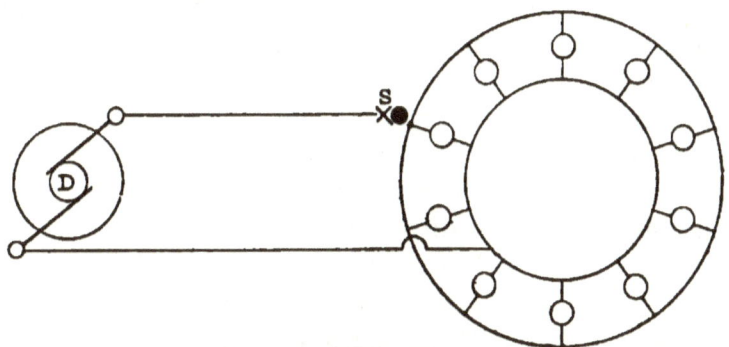

Fig. 52.—Diagram of Closed Loop, Parallel System.

dividual lamp switches) and safety fuses, or cut-outs, are placed upon a general switch-board, to which the mains from the dynamo are attached.

This switch-board is usually enclosed, under lock and key, and is called a "distributing box." From this point radiate all the circuits, with safety fuses at their roots, also double cut-out switches. The safety fuses should be fitted to both negative and positive wires. On the alternating current system there is no negative and no positive wire. Each wire becomes a — and a + pole many times in a second. American wiresmen call this box a "closet."

Before entering upon the uses of cut-outs, fuses, and so forth, it may be as well to point out that there is a certain advantage in some cases in locating all the accessories of the circuits at one point. It appears especially suitable to hotels and large institutions. On the other hand, it has its disadvantages.

Closed Loop Circuit.—Lamps are very frequently put in a closed loop circuit, as in Fig. 52. In this case it will be found most advantageous to connect as shown, or if the loop be large, to connect feeders from opposite sides.

The Tree System.—Professor Forbes * has given the name "tree system" to the plan represented in the diagram Fig. 53. Here we have the connections to the mains—street or dynamo—with "main fuses" and key switches—at the root of the tree. Thence lead a pair of sub-mains, forming its stem, throughout the main length and breadth of the building. From them spring "branches" or room circuits, and from these "twigs" or single lamps, representing the leaves.

Thus there are three sizes of wires usually employed. Coarse wire for the house mains, medium for the branches, and finer for the twigs, according to the current to be passed by these wires.

It will be observed that the safety fuses in this case (each fuse being represented by a *black* circle, and the lamps by light circles) are distributed throughout the system, one at least being placed at the root of each branch. Keys, or switches, represented by ×, are also placed at the roots of the branches, or as near thereto as may be convenient. It is usual to fix switches and fuses close together.

* See "Cantor Lectures," given before the Society of Arts, Feb. 1885, by Prof. George Forbes.

SIZE OF WIRE FOR THE CIRCUITS. 143

The three-wire system, already described (p. 132), is sometimes used in order to reduce the pressure from the mains to one-half, and to effect a saving of conductors. But it is only in particularly extensive installations that this would be done.

Size of Wire for the Circuits.

The Board of Trade rule allows a current of 2000 ampères per square inch of section of pure copper conductor of equal conductivity. This current will, however, be found

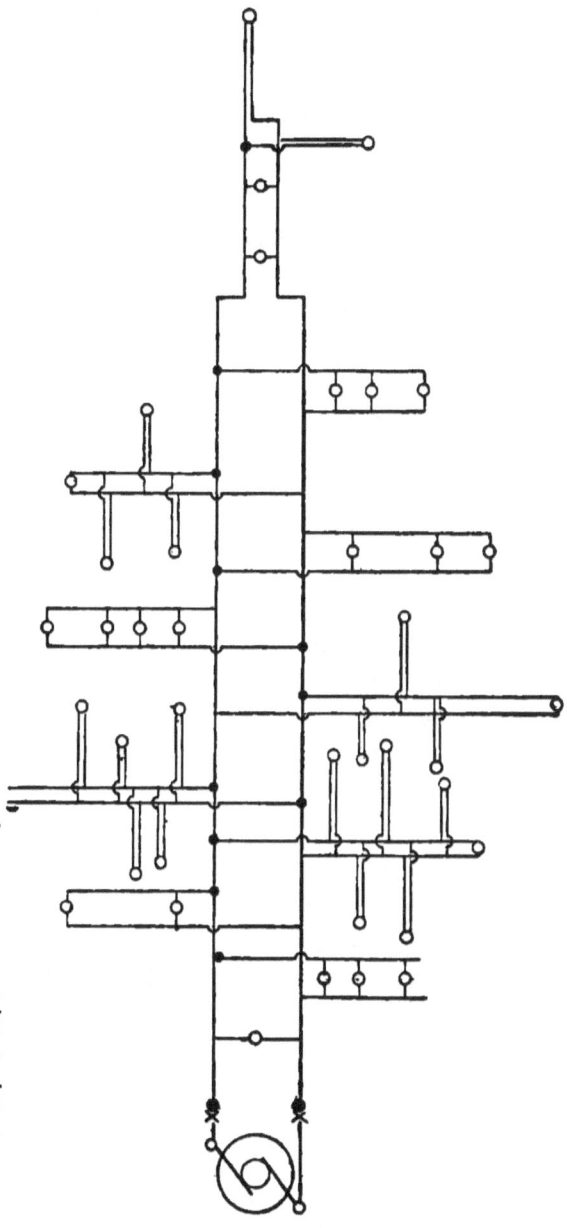

Fig. 53.—Diagram of the Tree System.

to make common copper wire rather hot. Practical electricians seldom allow more than half this current in the conductors — 1000 ampères per square inch. This current will not sensibly warm a wire of 95 per cent. conductivity, and is quite safe.

According to this latter rule it is only necessary to consult a reliable table, giving the standard sizes of wires, with their *section per square inch*, to ascertain the gauge required to carry a current of so many ampères.

It is a common practice in works on electric lighting to provide a mass of untested data from various sources. The beginner is expected to puzzle over these as best he can; but it will be found that, save in rare instances, such information is unfitted for the use of practical men. We do not propose to follow this rule, or at most will give only one or two examples of useful tried formulæ essential in practical work. Nor do we propose to weary the reader with cut-and-dried "examples" and "demonstrations" of the formulæ, for the simple reason that they are of no utility to a man about to plan an electric wiring system based upon conditions whch can never be foretold in a book.

The following table provides in a ready form a good deal of information. The first column gives the sizes of wires in use in this country, according to the legal standard gauge brought into force recently. This gauge is very similar, save in fine and coarse sizes, to the older Birmingham wire gauge. When a conductor is as thick as a lead pencil No. 6) it is too stout and stiff to be used for wiring. The practice is to substitute a *cable* composed of several smaller wires stranded. Columns III., IV., V., and VI. are generally useful, while columns VII. to X. are of especial importance to the electrician. Column IX., giving the sectional area of the wire in square inches, enables a rapid calculation to be made as to the required size, on the generally used basis that 1000 ampères of current can be allowed per square inch area. Columns XII. and XIII. give the resistances in ohms per 1000 feet and per pound weight, figures which will be found a useful check upon the quality (conductivity) of the wire, and in testing.

Column XIV. is based upon the rule which allows 1000 ampères per square inch, and is approximately correct—for practical working quite safe. Many successful electricians work to such a column as this with perfect satisfaction.

Columns XV. and XVI. are approximate only, and give the number of lamps of different voltage that are usually successfully run from the wires.

The figures given in the table are intended to apply more especially to house wiring, where the distance between the dynamo or mains and the furthest lamp does not exceed 100 yards. It is taken for granted that not more than 50 lamps are placed upon one circuit.

In taking resistances of circuits it must be borne in mind that the resistance of a parallel system of electric lamps and leading wires is a combined resistance, of which the component parts are the resistance of the leading wires and the resistance of the lamp filaments. The working resistance of a lamp is only to be got when it is lighted up. Its resistance cold is much greater than this.

If a pair of wires be set out, and one lamp be put

TABLE NO. II. FOR THE USE OF INCANDESCENT WIRESMEN.

I	II	III	IV	V	VI	VII	VIII	IX	X	XI	XII	XIII	XIV	XV	XVI
Standard wire gauge.	Number of wires (if stranded).	Diameter.				Equivalent to solid wires.				Length and weight.	Weight and resistance.		Safe working current on basis of 1000 ampères per sq. in.	Approximate number of lamps usually run on the wires.	
		Of each single wire.		Of the strand.		Diameter.		Sectional area.		Pounds per 1000 ft.	Ohms per 1000 ft.	Ohms per lb.	Ampères.	45 to 60 volt lamps.	90 to 110 volt lamps.
		In.	m/m.	In.	m/m.	In.	m/m.	Sq. in.	Sq. m/m.						
22	1	·028	·711	—	—	·028	·711	·0006	0·397	2·37	13·167	5·54848	0·8 to 1·0	—	1
21	1	·032	·813	—	—	·032	·813	·0008	0·518	3·10	10·081	3·25229	1·3 ,, 1·5	1	1
20	1	·036	·914	—	—	·036	·914	·0010	0·656	3·71	8·427	2·27254	1·5 ,, 2·0	1	2 to 3
19	1	·040	1·02	—	—	·040	1·02	·0012	0·810	5·34	5·852	1·09596	2·0 ,, 2·5	2	3 ,, 4
18	1	·048	1·22	—	—	·048	1·22	·0018	1·167	7·27	4·299	·59157	2·5 ,, 3·0	2 to 3	4 ,, 5
17	1	·056	1·42	—	—	·056	1·42	·0024	1·588	10·17	3·069	·30135	3·0 ,, 3·5	3	5 ,, 6
16	1	·064	1·62	—	—	·064	1·62	·0032	2·075	12·79	2·443	·19104	3·5 ,, 4·5	3 ,, 4	6 ,, 7
15	1	·072	1·83	—	—	·072	1·83	·0040	2·626	15·69	1·991	·12679	4·0 ,, 4·5	4	7 ,, 8
14	1	·080	2·03	—	—	·080	2·03	·0050	3·242	20·85	1·498	·07186	5·0 ,, 5·5	4 ,, 6	8 ,, 9
13	1	·092	2·34	—	—	·092	2·34	·0066	4·287	27·32	1·144	·04187	6·0 ,, 7·0	5 ,, 7	10 ,, 12
12	1	·104	2·64	—	—	·104	2·64	·0085	5·480	35·96	·869	·02416	8·0 ,, 9·0	6 ,, 8	14 ,, 16
11	1	·116	2·94	—	—	·116	2·94	·0105	6·774	43·59	·717	·01645	10·0 ,, 11·5	7 ,, 10	16 ,, 18
10	1	·128	3·25	—	—	·128	3·25	·0128	8·302	54·35	·575	·01058	12·0 ,, 13·5	9 ,, 11	18 ,, 20
9	1	·144	3·65	—	—	·144	3·65	·0162	10·50	66·30	·471	·00711	15·0 ,, 17·5	10 ,, 14	24 ,, 28
8	1	·160	4·06	—	—	·160	4·06	·0201	12·97	82·41	·379	·00460	20·0 ,, 22·5	12 ,, 20	35 ,, 38
25	3	·020	·508	·042	1·07	·034	·863	·0009	0·585	—	—	—	—	16	—
23	3	·024	·609	·051	1·29	·042	1·006	·0014	0·893	—	—	—	—	—	—
22	3	·028	·711	·059	1·50	·049	1·24	·0019	1·216	—	—	—	—	—	—
25	7	·02	·508	·060	1·54	·053	1·35	·0022	1·423	—	—	—	—	—	—
23	7	·024	·609	·072	1·83	·064	1·62	·0032	2·075	—	—	—	—	—	—
22	7	·028	·711	·084	2·13	·075	1·90	·0044	2·849	—	—	—	—	—	—

RESISTANCE OF THE CIRCUIT. 147

across their far ends, the resistance of that circuit will be that of the wire added to that of the lamp. If another lamp be placed across the wires, the resistance of the circuit falls considerably, because a fresh additional path has been opened to the current. If three lamps be used it falls still more. The greater the number of lamps across the wires the less the resistance. In this way the resistance due to lamps is easily obtained by dividing the resistance of one lamp (in ohms) by the number of lamps.

$$R_3 = \frac{\text{Resistance of a single lamp, hot.}}{\text{Number of lamps in parallel circuit.}}$$

In *series-parallel* the resistances of the lamps will nearly follow the same rule. If two lamps in series are to cross the mains, they may be treated as one lamp with their resistances added together.

Wire Gauging.—The gauge of a wire is an important point. If it vary from the dimensions calculated for, it may easily lead the wiresman astray. The necessity for actual measurement as a check upon the reputed gauge of a wire has been of late forced upon the attention of engineers. There is now in this country but one table of gauges—that authorised by the Board of Trade and known as the standard wire gauge, adopted, as to the required numbers, in the tables given in this book.

It is very convenient to carry a pocket gauge of sufficient range and *accuracy* to cover the requirements of ordinary wiring. Several of these have of late been introduced, to meet a demand which is doubtless increasing. One of the best of the improved guages is represented in Figs. 54 and 55, which shows the actual size of a very portable and accurate form patented by Mr. Trotter. This little gauge is provided with four scales. The standard sizes are given on the scale marked *S.W.G.* The scales marked *inch* and *millimetres* give the diameters of a wire in decimals of an inch and millimetres respectively, both being furnished with verniers. Each has one scale with an arrow head and one without. The latter is the scale proper, the former is the ver-

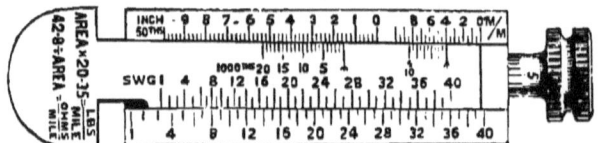

Fig. 54.—Trotter's Wire Gauge. Back.

nier. The arrow head points to the graduation on the scale from which the approximate reading is taken. The first decimal figure is read on the scale by the direct indication of the arrow head. It will then be found that one of the graduations of the vernier coincides with one of the graduations of the scale, and the remaining figures required to complete the reading

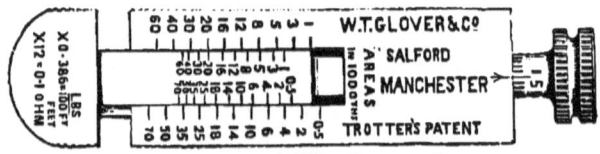

Fig. 55.—Trotter's Wire Gauge. Front.

are the numbers which correspond to this graduation, counting from the arrow head. The instrument is opened by turning the screw to the right. The wire to be measured is placed between the jaws and nipped tightly. The area of circles may be read upon a scale upon the back of the gauge. The area of a wire, once known, gives at once the capacity of that wire for carrying a current. It is understood that high conductivity copper wire is alluded to, and that some such constant as 1000 ampères per square inch sectional area is used.

The gauge represented in Fig. 56 is an ordinary pocket hole or gap gauge, provided with apertures for all the usual sizes employed in wiring. It may be well to note that standard gauge copper wire, which is afterwards *tinned*, will read a fraction of an inch larger than standard gauge. Before measuring a wire all insulation should be carefully removed.

Tests for Conductivity of the Wires.—In fitting up a

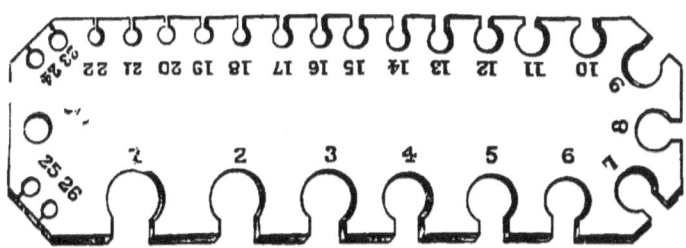

Fig. 56.—Gap Wire Guage.

large installation specimens of the conductors to be used should be tested for both conductivity and resistance. Each sample should not be less than 100 ft. in length. Each length, if intended for damp situations, should be immersed in a tub of water for at least 12 hours. The wires should then be tested by the aid of the testing box and figures given at p. 76. The resistance of each length should not exceed that given opposite to its gauge in the preceding table. The insulation resistance must depend upon the nature of the covering. If the wires are "best" insulated, in gutta-percha and tapes, they should show an insulation resistance of 20 meg-ohms per length.

Nature of the Insulating Covering.—The commonest insulated wire that can be safely used for electric light branch wires is coated as follows :—

(1) Tinned copper wire, conductivity 95 per cent. One coating india-rubber; braided with cotton and coated with preservative compound. Such a wire is unfitted for immersion in water, or for work in damp situations.

(2) Tinned copper. One coating cotton; one coating india-rubber; one coating felt; braided cotton and preservative compound. Such a wire is adapted for more exposed work.

(3) Tinned copper. One coating cotton, saturated with paraffin wax; one coating pure india-rubber; another coating cotton tape; braided and coated with preservative compound. This is "good insulation" adapted for damp situations.

(4) The same as above, with two coatings india-rubber.

(5) Vulcanised insulation, consisting of vulcanising india-rubber; one coating rubber-covered tape, and the whole vulcanised together and coated with preservative compound.

(6) The same as above, with a heavy braiding over all.

(7) Highest class. As above, with a covering of *lead* over all.

(8) Twin-wires, consisting of several fine wires, *e.g.*, 100 No. 40 stranded together, insulated separately, braided in pairs, for flexible conductors. These are used for portable lamps. The exterior covering is in silk, mohair, or cotton.

One rule should be followed by the electrician responsible for the success of the wiring—to use the best class of insulation the nature of the case will permit, and to avoid having the wires too fine.

Switching Arrangements.

In planning the wiring of a building the main and lamp switches should be marked off on the plan. It is impossible to say how many switches should be fitted to a given number of lamps, unless the conditions be known. In house wiring it is convenient to provide a double main switch on the house side of the meter, and a switch to each lamp fitted. In electrolier work, where all the lights would be required at one time, a single switch for that group will answer.

It is neither necessary nor usual, as with gas, to fix the switches close to the lamps, especially when these are overhead. It is more convenient to furnish the means for lighting and extinguishing either near the doorway, as in hotel bedrooms, or near to the fireplace, as in drawing-rooms. Each case must be made to decide for itself. There is one point, however, that is worth considering carefully; it is a great saving of labour to locate switches and cut-outs (safety fuses) together; and it is an advantage to keep the fuse as near as possible to the root of the wire supplying the lamp.

Main Switches.—These are fixed at the root of the system, or at the root of each branch circuit, according to the nature of the case. There are many patterns in use, and we can only notice one or two of them. One of the most efficient of the double break type is represented in Fig. 57, which shows Drake & Gorham's main double ring-contact switch. The cross-arm, provided with insulated handles, swings upon the central staff. The contact between it and the terminal rings shown is effected by slitting the ring, so providing a spring clip, into which the curved end of the cross-arm enters. Fig. 58 will serve to make this

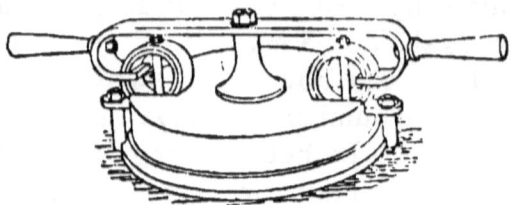

Fig. 57.—Drake & Gorham's Ring Contact Switch.

clearer. The ring is made in two or more parts, which gives it elasticity, and certainty of contact, as it wears away, is kept up by adjusting the lock nuts.

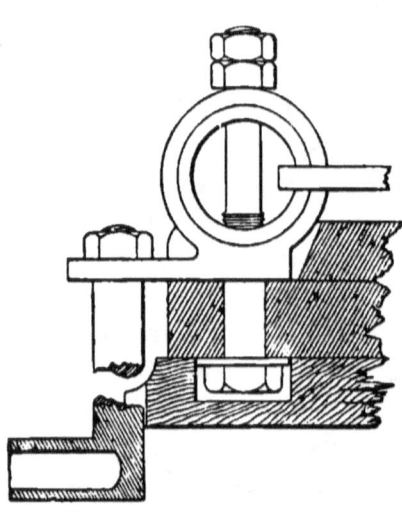

Fig. 58.—Ring Contact, Section.

This method of getting a tight sliding contact of large surface has been applied by the same inventors to a large variety of switches, some of which are represented upon the main switch-boards illustrated further on. Such switches are invariably mounted upon incombustible bases, of which perhaps the best is slate. Their main function is to provide perfect continuity when closed, and perfect safety from creating an arc, and so setting up a fire, when opened. Fig. 59 represents one form of Woodhouse & Rawson's double-break switch.

Double-pole main switches are different from double-break switches inasmuch as they are constructed in duplicate—two switches in one, so that by one movement of the handle both leading and return wire are cut off from the circuit. Instead of opening the two wires of one circuit a double-pole switch may be used to open the positive or negative wires of two separate circuits.

Of late a good deal of objection has been raised by

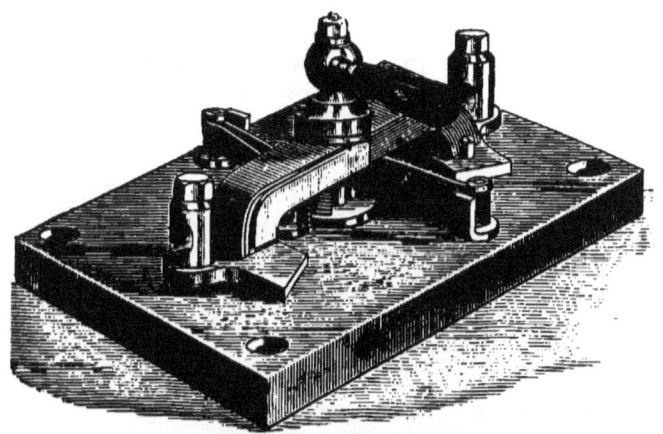

Fig. 59.—Woodhouse & Rawson's Double-break Switch.

fire office authorities and others against the practice of carrying wires belonging to the same circuit so close together, as is frequently necessitated by the use of double-pole switches. Any objection of that kind may, however, be easily met by providing two single switches, one upon each wire, a sufficient distance (several inches) apart, or by the simple expedient of providing sufficient space upon the double-pole switch itself. There can be no doubt that a double-pole switch is a great convenience. Fig. 60 represents Mr. Hedges' device for this purpose, which is now extensively used in house lighting in the metropolis. A and A' are a pair of sprung contact discs, bearing with sufficient pressure upon the polar pieces B B'. The base is incombustible. The main cable connections are made into the set-screw sockets as

Fig. 60.—Hedges' Double Pole Switch.

shown at + —, and the branches from the opposite side.

Multiple-Way Main Switches.—Main switches may be broadly regarded as effecting three changes: (1) Cutting open and closing one wire of a circuit—or, in other words, "off and on" switching; this class of single-wire switch is made both with a "single break" and a "double break," as shown in

Fig. 59. The double break divides the spark at breaking circuit, and will suffer less than a single-

Fig. 61.—Woodhouse & Rawson's Multiple-way Switch.

break switch from burning of the contact surfaces. (2) Cutting open and closing the two wires of a circuit,

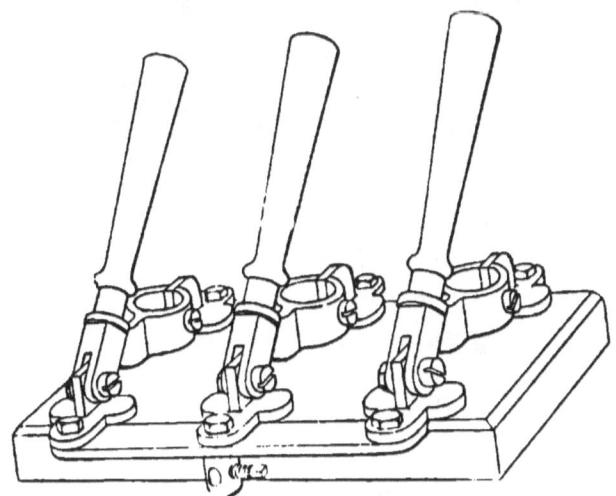

Fig. 62.—Ring-contact Multiple-way Switch.

called a *double-pole* switch. This variety severs the mains completely from the branches. The importance of this will be discerned when it is pointed out that in the event of a bad *short circuit to earth* in one of the branch wires, merely severing one wire rom the mains would not necessarily stop the leakage, while severing the two cuts off all possibility of main current reaching the branches. (3) Multiple-way, or distributing switches. These are arranged in a variety of ways. The simplest form is represented by a central contact connected to the positive main, having a lever moving at will on to any one of several branch line

Fig. 63.—Woodhouse & Rawson's Accumulator Switch.

terminals, so as to throw the main current into any required wire. The switch may of course be elaborated to any degree, so as to lead the main current into any number of wires, according to the nature of the case. Figs. 61 and 62 show two forms of this description of switch, in the first of which Woodhouse and Rawson's multiple contact plate system is employed; the contact being given by a number of springy slips of gun-metal or brass, an arrangement which ensures a good deal of bearing surface between the two contacts. Fig. 63 is a form used for placing accumulators in and out of circuit during charge or discharge. The short-circuiting of a cell is obviated by a coil of wire placed upon a vulcanised fibre plate below the slate base. It may be pointed out that if a switch carrying a heavy current have but a small contact, great heat and ultimate burning may be set up at that point. In the selection of switches, especially main switches, only those furnished with incombustible bases of sufficient thickness to retard the passage of heat should be used.

Branch Line and Lamp Switch.—These have been produced in great variety. The simplest kind in use merely forms a metallic touch, connected to the leading wire, and so arranged as to throw the current into the branch by contact. This is by far the most common variety of switch. It frequently is found with a great defect, which is worth careful consideration before fitting switches to be manipulated by ordinary people. A common switch, if turned partially but not wholly off, may serve to extinguish the lamps, but at the same time may be in partial contact sufficient to set up an arc there, or a fusing heat. This is he worst kind of switch possible for fitting into nouses. The difficulty can be overcome by fitting the movable lever with an "overthrow" spring, the function of which is to rapidly push open the switch as soon as it is started by hand, a device which renders "arc-ing" impossible. This re-acting spring device has recently been considerably improved upon in the Woodhouse & Rawson switches, and in those of several other makers. An "overthrow" spring may possibly be resisted by the hand of the person opening the switch long enough to form an arc and burn the switch. Such a contingency is met by making the

158   WIRING FOR INCANDESCENT LAMPS.

little handle free upon its arbor, so that when the switch is started once it is thrown fully open by the spring independently of the handle; *e.g.*, even if the handle be held firmly after starting the switch, it will spring open and prevent " arc-ing."

These switches are furnished with channels for the connection of the wires from the back, and with the

Fig. 64.—Branch Line or Lamp Switch.

necessary screw-holes for their fixing. They are usually protected by a cover of porcelain, or wood, of a shade and style of ornamentation to suit the colouring of the apartment in which they are placed. Fig. 64.

Combined Switch and Cut-out Fuse.—A very useful form of switch is that in which a fuse is fitted, as represented in the front of Fig. 65. The fuse usually

COMBINED SWITCH AND CUT-OUT. 159

consists of a slip of tin, or alloy, which is easily re-

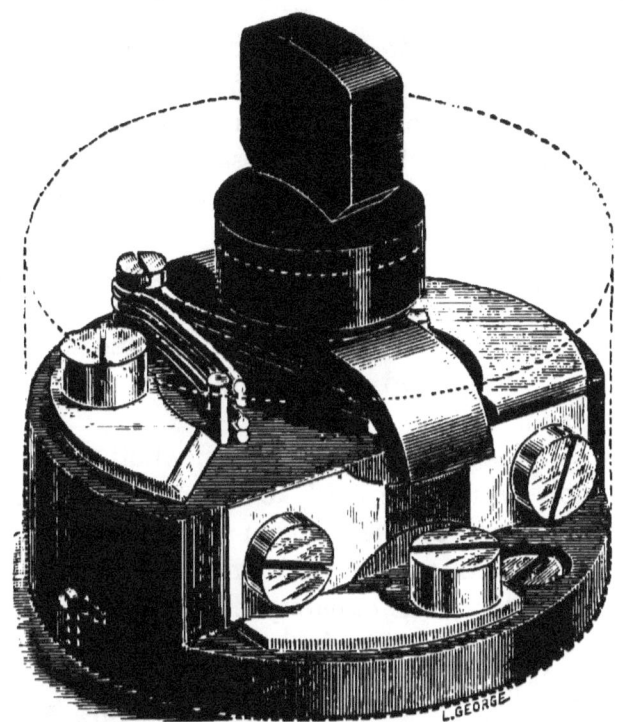

Fig. 65.—Switch and Fuse combined.

placed when accidentally burnt by too heavy a current. These combined switches are likely to come into general favour, reducing as they do the labour of fitting fuses.

When a wall plug is provided for the flexible leading wires of a portable lamp, a duplex plug connection, as represented both in

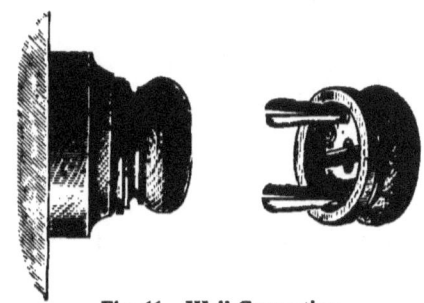

Fig. 66.—Wall Connection.

position and separated in Fig. 66, is generally employed.

Plug and Removable Key Switches.—A plug switch is merely a metal plug, as used in resistance coils, which is inserted or withdrawn to make and break contact. Key switches are those opened and closed by a separate key, which is easily removed and carried about.

Reversing Switch.—The ingenious device shown in Fig. 67, which is due to a French electrician, enables the switching of a lamp to be effected from two points, as either end of a room. The diagram explains itself.

Capacity of the Switch.—Switches, whether for main or branch work, are said to be made to carry so many ampères—in the case of branch switches this would be approximately as so many lamps. A wide margin of carrying capacity is generally allowed. The main provision to insist upon is perfect contact, insulation, and incombustibility.

Main Fuses and Cut-outs. — A safety fuse is a device to prevent an accidental abnormal current from forming in a circuit. If a current greater than the circuits in a house were designed to carry were to pass through them, the points offering greatest resistance would speedily become red-hot, and fire would probably ensue. The main object, then, of a safety fuse, or cut-out as it is frequently called, is to prevent accidental overheating.

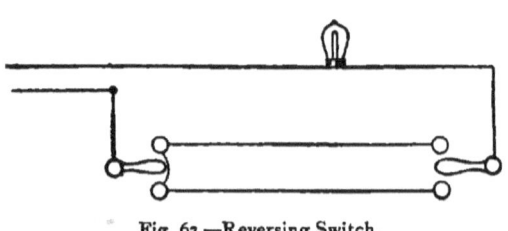

Fig 67.—Reversing Switch.

When a small portion of a circuit is composed of *very fine* copper wire, it will break at that point as soon as the current is raised sufficiently high to melt the copper. Perhaps the best material for a fuse is copper, but it has one great objection—its high fusing point. It is usual to employ an inch, or thereabouts, of tin, or tin-lead alloy wire; and in respect to gauge, it is a common practice to employ a wire one size finer than the copper circuit wire—*e.g.*, a 16-gauge copper wire

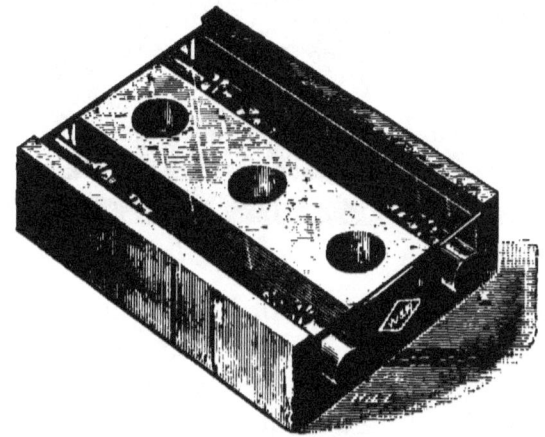

Fig. 68.—Main Fuses.

would be protected by an 18-gauge tin wire. Main fuses inserted at the root of house mains take several forms. One of the most convenient is the W. & R. double-pole fuse, represented in Fig. 68, consisting of a slate double trough so fitted that it may be conveniently screwed to wall or other fixture. The wires are led to the metallic plugs, which are fitted with screws for their reception. Between the plugs short lengths of lead or tin wire are fixed, and the whole is covered by a glass cover. Thus any accident to the fuses can be detected and the fusible wire replaced.

The capacity of the fusible plate is usually indicated upon it in ampères as shown in the diagram (Fig. 69) of Hedges' main fuse. The fuse plate may easily be removed and replaced by others (Fig. 70).

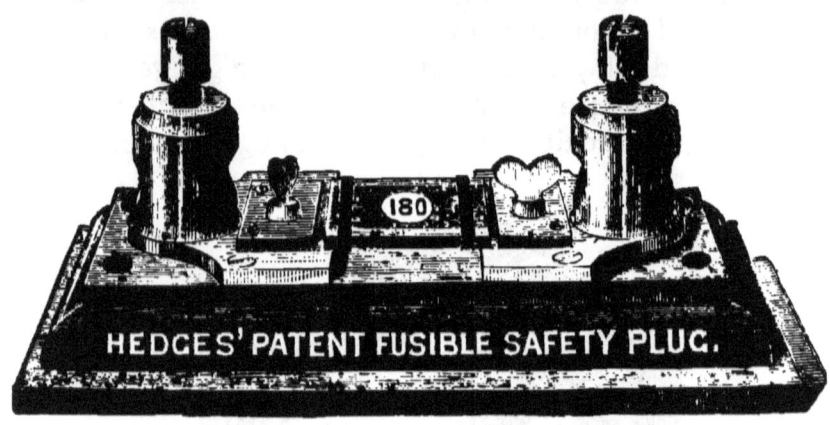

Fig. 69.

In many systems plugs of lead or fusible alloy are used for this purpose; in others merely a loop of lead wire is employed. There is one certainty in the use of a good fuse : if it be fixed close to a dynamo it will always burn up long before the copper wires are hot enough to injure the insulation of the machine. A paper by W. H. Preece, F.R.S.,* gives several deductions from experiments upon the fusing of different metals under the current.

Fig. 70.—Safety Fuse Plate.

The following refers to copper and tin lead alloy, two substances very much used for fuses :—

* Proceedings of the Royal Society. Vol. XLIV. March, 1888.

BRANCH WIRE FUSES.

Standard wire gauge.	Copper wire fuses at ampères.	Tin-lead alloy wire fuses at ampères.
14	231·8	29·82
16	165·8	21·34
18	107·7	13·86
20	69·97	9·002
22	48·00	6·175
24	33·43	4·300
26	24·74	3·183
28	18·44	2·373
30	14·15	1·820
32	11·50	1·479

The fusible alloys generally employed for making safety plugs are usually more strictly amalgams in the case of the softer plugs. These are made from Arcet's metal, 9 parts, mercury, 1; and fuse at about 50° C. Harder plugs, melting at 210° Fahr. (just under the heat of boiling water) are made of tin, 3; lead, 5; bismuth, 7. A useful fuse for wires hung bare, where a good deal of heat may be allowed with safety, is made from tin, 1; bismuth, 1; it fuses at 285° Fahr.

Fusible Plugs and Branch Fuses.—The usual safety plugs are marked with the *number of ampères* of current they can carry without fusion. They are also, in rare cases, marked in "lamps," but this practice is extremely misleading.

Fig. 71.—Branch Fuse.

If marked in lamps, 16 candle lamps will probably be meant in most cases. Since the function of a safety plug is to protect the circuit at whose root it is fixed, it has no reference to lamps. The ampères in that circuit may approximately equal the number of lamps, or they may not, according to the voltage of the lamps (*e.g.*, 50-volt lamps might take ampère, while 100-volt lamps would take ·5 ampère). One of the simplest forms of fuse plate is represented in Fig. 71, where milled nuts are provided for replacing a burned out connection.

Messrs. Patterson & Cooper issue a convenient and substantial form of fuse upon a slate base, represented in Fig. 72, in which a flat foil is used as in Hedges' fuses.

Particular Observation Respecting Fuses.—The inser-

Fig. 72.—Main Fuse Plate.

tion of a number of fuses into a system of circuits may render that system a very safe one as far as overheating by accidental abnormal currents is concerned. But the multiplication of fuses may easily become a source of danger instead of safety. Each fuse implies a break in the wire and a pair of connections. Unless the connections are honestly thorough they become a source of trouble. Every fresh connection is a fresh weak point in the circuit. It is essential to so make the connections that each has a carrying capacity greater than the wire itself, and is unquestionably sound. A carelessly soldered connection may heat very quickly. It may at any time break apart and set up an arc, so igniting dry woodwork. Besides the connections the plugs them-

CAUTION RESPECTING FUSES.   165

selves, if plugs be used, may have bad connection in their sockets. Every plug should be examined previously to being inserted into the circuit.

Fuse Boards. — For reasons stated above many electricians will not permit the distribution of fuses throughout the system of house wiring. In such cases it is considered safer to assemble all the fuses upon a fuse board (Fig. 73). The base or backing of this "board" is of slate, and the terminal blocks of brass. The figure represents a fuse board for eight circuits. The connection to the positive main is made with the central screw, and the current is thus distributed into eight paths. The fuses themselves, shown by the light lines between the lower and upper terminals, are either of lead wire or of strips of alloy.

Fig. 73.—Fuse Board.

Fig. 74.—Terminal Block.

It is much easier to ensure good connections upon such a grouped system of fuses than in a distributed plan. But the use of fusible wire is difficult and uncertain, unless precautions be taken. The wire must either fit the terminal blocks very tightly, so as to ensure good contact without much screw pressure, or a piece of copper wire must be soldered to the end of the fuse for connection with the terminals. Although the fuse boards are usually placed only in the positive main, with a plain "terminal block" (Fig. 74) upon the return, it is still better to have a fuse board upon both poles of the main. The use of terminal blocks of this pattern is extending. They obviate the necessity for large, coarse main or ter-

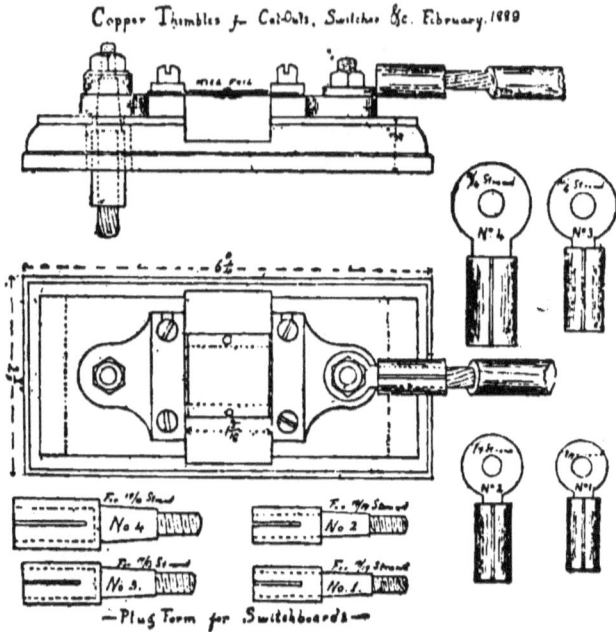

Fig. 75.—Hedges' Fuses for Switch-boards.

minal joints, and greatly facilitate examination and testing of the circuits.

Mr. Hedges has devised a form of fuse for switch boards (Fig. 75), which would appear to offer several advantages, providing as it does means of rapidly connecting the main cables. The fusible foil bridges over a gap as represented.

Mr. Scott has devised the handy fuse represented in Fig. 76. It consists of a fine wire, run over the surface of insulative material, forming a kind of conductive plug which may easily be slipped into position in the base. This plug provides the " one inch break " generally required by the fire offices.

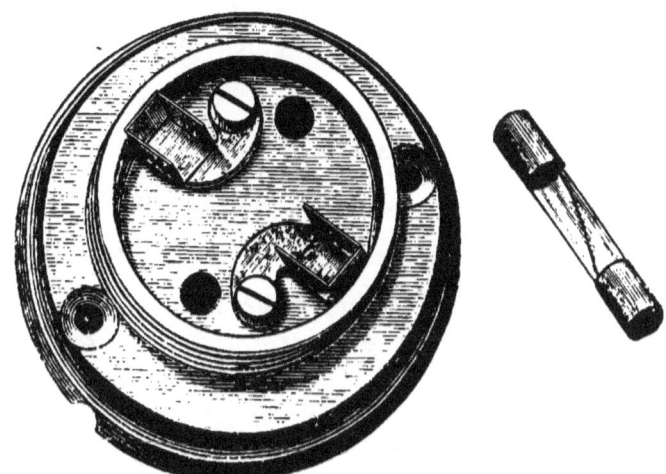

Fig. 76.—Scott's Fusible Plug.

The Lamps and Fittings.

Area Lighted.—A 16 candle-power lamp is usually fixed for every 100 square feet around the lamp, the latter being raised from 6 to 10 feet above the floor line.

Light Absorbed by Glass Envelopes.—When the lamp is covered by a globe the following percentages of light are lost for the different classes of glass: Clear glass, 10 p. c.; ground glass, 35 p. c.; opalescent, 50 p. c.

Incandescent Lamps in General Use. — Various makers' productions are in use. The most widely known are the Edison & Swan's lamps. They range, according to the nature of the filament, from one candle power to 1000 or more, when fully incandesced. The most common powers are the 8, 10, 16 and 20 c. p. lamps. For house lighting 10 and 16 c. p. lamps are deemed sufficient. Of the two the 16 c. p. is probably the more widely used.

Electromotive Force and Current of the Various Classes.—The lamps in general use absorb approximately the following:—

		Volts.	Ampères.		Volts.	Ampères.
8 candle-power takes from	10	..	2·8	to	120	..	0·3
16 ,, ,, ,,	15	..	3·7	,,	160	..	0·4
25 ,, ,, ,,	40	..	2·2	,,	120	..	0·7
50 ,, ,, ,,	50	..	3·5	,,	120	..	1·4
100 ,, ,, ,,	50	..	7	,,	120	..	2·9

But it is impossible at the present time to lay down a rigid rule. The 8 c. p. lamps are very generally run at 10 c. p., and the 16 c. p. at 20. It is more economical, as far as current is concerned, to run them above their nominal value. But the life of the lamp is thereby shortened. The "life" varies, and depends almost entirely upon the supply—its constancy, regularity, freedom from fluctuations and so on. The average has been placed at 1000 hours, but many lamps have been known to burn 5000 hours. The life chiefly depends, no doubt, inversely upon the *intensity* of the incandescence.

Nature and Description.—The incandescent lamp is now so well known that it appears unnecessary to describe it. It may be regarded, however, as a short length of very fine conducting filament of graphite carbon, usually curved into the shape of a hairpin, and mounted within a pear-shaped glass envelope, from which the air has been very carefully exhausted. The two ends of the carbon thread are put into communication with the exterior of the glass by two fine wires of platinum, sealed into the glass (Figs. 77 and

BLACKENING OF THE BULBS.

78). The lamps need careful handling, as the carbon filament is very brittle and easily broken. If the glass be broken, the lamp is destroyed. If this rupture should occur while the current is passing through the lamp, the carbon thread will be at once consumed. The filament is caused to glow by a small portion of current being impelled through it under the "pressure" or electromotive force set up by the dynamo. The higher this pressure the less current

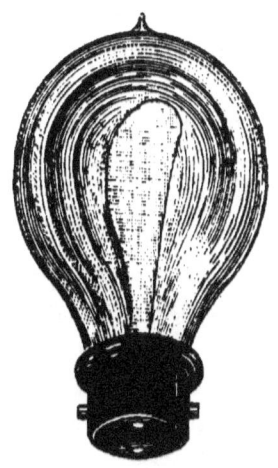
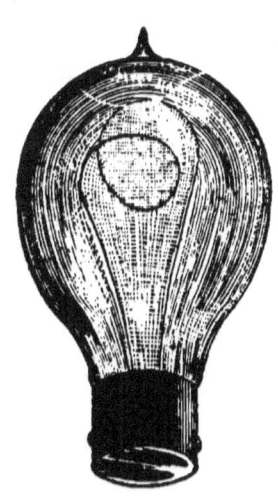

Fig. 77.—Edison Lamp, B.C. pattern.   Fig. 78.—Swan's Lamp, B.C. pattern.

per candle power is required, so that it is economical, as before stated, within reasonable limits to work lamps at a high pressure. The working pressure, measured in volts, is always marked upon the lamps. But, as in the case of various other "nominal" working figures, these are frequently exceeded.

Blackening of the Bulbs.—This cannot at present be avoided. When a lamp has been in use a few hundred hours the interior of the glass appears to get coated with a fine black powder—probably, particles of carbon from the gradual destruction of the filament. This blackening impedes the light, and it becomes a question whether it is economical to run such blackened bulbs longer after a certain percentage of light has been so cut off.

Current Absorbed by the Lamps.—The current varies, as before stated, with the "pressure" at which the lamps are worked. It is usually expressed in "watts." A lamp is said to take so many watts per candle power. The best lamps take from 3·5 to 4 watts per candle, under ordinary pressures. 746 watts are called an electrical horse-power, and from 45 to 60 watts will be absorbed by an average 16 c. p. lamp. Roughly, from 10 to 14 of such lamps are usually obtained per mechanical horse power expended upon the dynamo. The balance is usually lost in the resistance of the circuits.

The Economical Efficiency of the Lamps.—With regard to the question, *at what pressure it is economical to run the lamps*, seeing that a high pressure shortens their life, but calls for less electricity per candle power, an interesting paper was read at the American Institute of Electrical Engineers by Mr. Howell, electrician to the Edison Lamp Company, April, 1888.

The paper embraced a series of curves, showing the performance of lamps of various costs, candle powers, efficiencies, and with various periods of life. The general deduction from the numerous experiments that had been made to determine this point was that *the most economical efficiency of the lamps was attained when the cost of lamps was 15 per cent. of the cost of operating the entire electric plant.* In other words, if the lamp bills (renewals) were less than 15 per cent. of the total expense of the electric lighting, the pressure

THE KILOWATT. 171

imposed upon the lamps was too low. If the lamp bills exceeded 15 per cent. of the expenses, the pressure used was too high. It was also shown that if, for example, the lamp bills are only 10 per cent. of the whole cost, increasing the efficiency of the lamps by increasing their candle power does not reduce the total cost; but in order to attain that end the lamps must be replaced by others of the same candle power, but of higher efficiency. It is therefore clear that it is by no means economical to run lamps at so low a candle power that they will last beyond a certain number of hours. Instances are common of lamps having been run at so low an efficiency that they have lasted 5000 hours. It would appear that in this case it would probably have been better to burn out five lamps, each lasting 1000 hours.

Apart from the question of current wasted upon a lamp in the above way, there is the inevitable blackening of the bulbs to be contended against, as already spoken of. When this proceeds a certain way it is better to replace the lamp, even although its filament may have a good length of life left in it. It should, in fact, be treated as a broken lamp.

A "kilowatt" is 1000 watts. The kilowatt is frequently taken as a *unit* in describing the power of a dynamo. Thus a dynamo will be described as a "10-unit machine," meaning a dynamo capable of causing an electrical flow of 10 kilo-watts (10,000 watts).

The Board of Trade kilowatt-hour is the recognised unit of measurement of electricity supplied to consumers. It means a kilowatt maintained for one hour. Its selling price in this country varies from 6d. to 1s.

Fittings.—These may be roughly divided into sockets, or lamp holders, or connectors; brackets, or arms for supporting the lamps and pendants, and electroliers.

The sockets depend upon the connections provided at the lamp bulbs. These are arranged variously. The most common is a pair of metallic studs, fixed to the stem by means of a brass collar, and known as B C lamps in the trade. A common plan is to provide two small loops at the bottom of the bulb, and to hook them to the terminals in the socket to make electrical connection; called bottom loop lamps, or B L. The sockets are also made in the form of adapters, for use upon existing gas fittings, called G. F. adapter supplies (see "Gas Fittings," p. 196). The "contact" is made certain in various ways. The earliest plan was that of loops, kept apart by a spiral spring, and later by means of a bayonet joint (Fig. 79). The usual present device is arranged to come into contact by simply screwing, the details of which we cannot enter upon here.

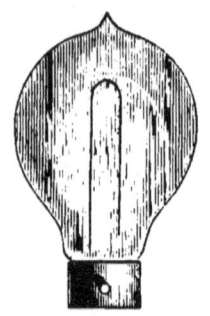

Fig. 79.
Edison Lamp, with bayonet joint.

Brackets are made in an immense variety of designs. They are very similar to gas brackets, but are generally arranged to throw the light downwards. Incandescent lamps being different from gas burners, inasmuch as they can be held in any position, afford great scope to the art-worker in the production of new and beautiful designs for brackets and electroliers. Flowers and fruit naturally suggest many ideas in this direction, and it is a common practice to make the lamp come in as a "bud," or the centre of a "bloom." or as the fruit itself. For such purposes the bulbs are frequently coloured.

A very convenient arrangement for students is represented in Fig. 80, which shows Mr. Hartnell's adjustable shade carrier, by means of which the light of the lamp may be projected in any direction, or at any angle.

Dispersion of the light is a subject which has occupied many minds, and there is probably nothing better

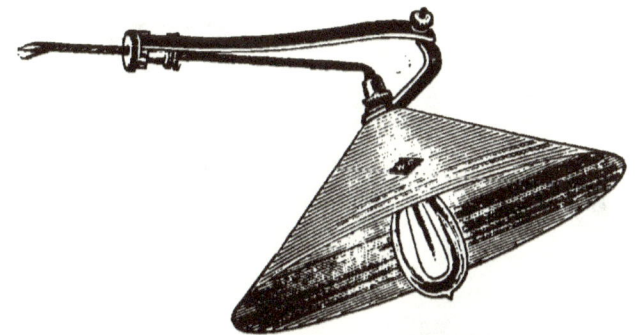

Fig. 80.—Hartnell's Lamp Reflector.

than reflectors for ordinary purposes. Mr. Trotter's application of dioptric shades appears to present some advantages in this direction. The shades are moulded from clear flint glass into innumerable little prisms, causing considerable diffusion of the light, while it obviates the glare of light direct to the eyesight. Fig. 81 represents one form of these shades, completely enclosing the lamp or lamps. The dioptric shades will no doubt prove useful for indoor arc lights.

Attachment for Portable Lamps.—Portable lamps, fed by a flexible twin wire, attached to a pair of poles fixed at any convenient point in the wall of a room,

174   WIRING FOR INCANDESCENT LAMPS.

are becoming very common. Since the leads are exposed to so much friction there is constant danger of short circuit in these leading wires. Hence the

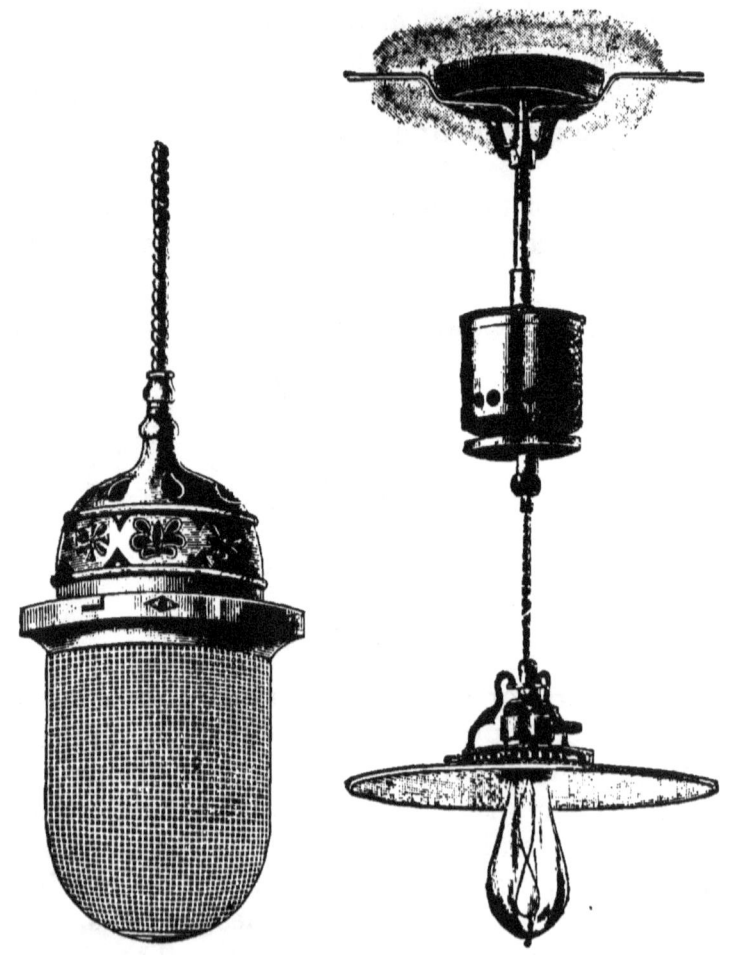

Fig. 81.—Trotter's Dioptric Shade.   Fig. 82.—Pendant Lamp, with Reflector.

invariable practice of careful engineers to fit a fuse behind the wall attachment, where the flexible leads leave the branch wires. If accidental contact then takes place between the leads, the fuse will give way before the danger extends to the wires themselves.

Telescope pendants have been devised for the electric light. Various plans have been suggested for he purpose of keeping up the *contacts* when the pendant is lowered or raised. One of the earliest ideas was to employ a flexible twin wire, running upon a spring reel, after the manner of a spring blind, the wire being used merely to feed the lamp. A later method employs a single sliding contact, consisting of a spring bearing upon an insulated metallic strip connected to the positive wire, the fitting itself being in contact with the negative pole. Various other devices have been tried which cannot be entered upon here. Fig. 82 represents a convenient reflecting pendant for either arc or incandescent lamps, issued by Messrs. Laing, Wharton, & Down.

Methods of Running the Wires.

"*Rule of the Road*" *for Leads.*—It may be as well to quote here the alliterative rule generally observed by wiresmen in running leading wires : " Leads left, Returns right," when laid upon a floor or ceiling; when placed upon a wall, horizontally, " Leads low, Returns raised."

Red insulation is generally used for leads (positive wires) and *Black* insulation for returns (negative wires). It may be observed here that a good deal of complaint is being raised that the *red insulation is inferior to the black*—this, if insisted on by makers, will speedily result in the black being used for leads, and the red for returns. For rules to find the direction of the current, see p. 24.

Cleat Wiring.—This means uncovered wires, run as neatly as possible upon walls, flooring, and ceiling, and held in place either by "cleats" (Fig. 83) of wood, with a double groove, or by leather loops.

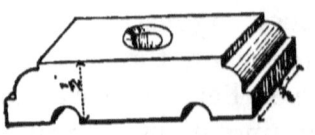

Fig. 83.—Double Wire Cleat.

Cleat wiring, although unfitted for house work, is eminently adapted for theatre stage-wiring, for mills, and in every situation where the appearance of the bare wires would not be objectionable.

It is *very desirable* to expose the wires to view if possible. It prevents moisture from accumulating, renders the detection of leakages and faults comparatively simple, and compels the wiresman to observe that the proper distance is maintained between the wires. In mills, where dust is generated, it is apt to settle *very much* upon wires carrying continuous currents, but not upon those carrying alternating currents. When a length of wire is run it should always be stretched taut, from point to point, and securely cleated down. Cleats may be required, according to the situation of the wires, from every three to every six feet of run.

Crossing Cleats have been devised for enabling one pair of wires to cross over or under another pair without danger of contact. They are usually made in glazed earthenware, but more frequently extemporised by the wiresman himself upon the spot. A wooden cleat of this kind merely consists of two cleats placed across each other. It is usual to put in a square of vulcanised rubber between the cleats. The single-wire crossing pieces are usually made in earthenware, and lead one wire, as an arch, over the other. The distance between the wires so crossed should *never be less* than an inch. The main provision is certainty of separation. It must be impossible to press or bend the wires so crossed into contact with each other.

Cleats should be screwed, not nailed. The screws must not touch the insulating covering of the wires.

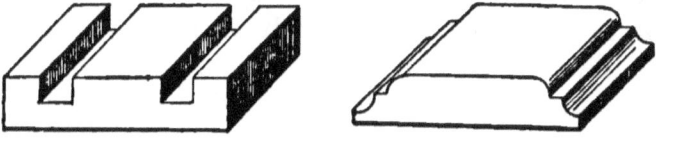

Fig. 84.—Double Wire Casing and Cover.

Brick walls must be pierced with chisel and hammer, to allow of the driving in of a block of wood upon which to screw the cleat. The wood should be driven in so that its grain is across the path of the screw, otherwise the latter may be easily pulled out. The cleats themselves are made from *hard* wood, with semicircular or square channels.

Casing and Moulding Wiring.—This is by far the most common method at the present time for house work. It implies concealed wires, but yet accessible in a case of need. The casings are merely continuous cleats. Fig. 84 is an example of a section of plain channelling with its moulded cover, and Fig. 85 of a more elaborate pattern. Figs. 86 and 87 represent single-wire mouldings for a cornice or angle and an open situation respectively.

Fig. 85.—Double Wire Moulding.

Casings are usually made in soft wood, but for

special purposes are produced in immense variety by Mr. Elliott, of Newbury, from whose designs the above engravings are taken. Mouldings will necessarily be selected to suit the ornamentation of the rooms, or to taste.

There are several methods of casing the wires. But all that is necessary is to run the wires taut from point to point, and to securely screw the moulding upon them. Some wiresmen are more particular in their method, and take pains to loop down the wires upon the walls first, the requisite distance apart, and to apply the casing merely as a covering or protection. The channelling, which is double, as Fig. 84,

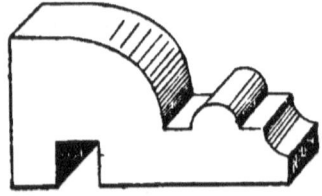

Fig. 86.—Cornice Moulding.

Fig. 87.—Single Wire Casing.

is first screwed to the walls or floors, and the cover laid upon it after the wires are run in the grooves.

The object of mouldings is of course to conceal the presence of wires altogether, and numerous ingenious devices have been resorted to by artistic workmen to get wires to fittings and electroliers without breaking the plaster of the walls. In some cases the plaster is cut out in the path of the wires, and, after they are laid in a thin sheathing in the channel so obtained, a thin wooden cover is put on, and the whole re-papered or painted, forming complete concealment. Mouldings are frequently run above the wainscot, or in corners, or along the course of skirting boards. When wires have to be run upon a ceiling, and the place for the lamp *cannot be reached from above*, a moulding must be run across; but it is usual in that case to give the ceiling a symmetrical appearance by fitting three other *dummy* mouldings, forming panels.

It is impossible to enter here upon the numberless devices resorted to for the purpose of concealing the wires, or at least giving their covers an artistic appearance. Each case must be made to decide for itself.

It is almost needless to enter upon a consideration of the wiring of buildings while in course of construction. Although the use of the electric light is spreading rapidly it will not for many years be allowed for in new buildings in this country, except in rare instances. Progress in England is extremely slow, and it is probable that houses will continue to be fitted for gas light long after that illuminant has been relegated in great part to the duties of heating and cooking.

If a general suggestion may be thrown out we may say that the architect should provide vertical tunnels in the brickwork, at least 8 in. by 12 in., communicating with each floor from top to bottom of the house. The tunnels should be boarded or wood-lined by the carpenters. That, with the provision of horizontal openings of the same size through partition walls, on the level of each floor, will form the only difference necessary between providing for gas and electricity. Ceilings will be reached from above, as in the case of gas-fitting. Brackets on walls will be reached from above or below by means of small tunnels formed behind the plaster. Gas-pipes are buried in the plaster, or cleated to the brickwork. This cannot be done in the case of electric light wires, and it is doubtful whether it should be resorted to, even in the case of wires covered over all by a protection of lead.

Wherever wires are run in a building the adjacent woodwork must be dry, and conductors must in no case be affixed to, or laid in damp walls.

In running wires beneath flooring, and in other situations where the wires cannot be cleated down, it is important to ensure that they are "hauled taut" and well separated; in running concealed wires this precaution against accidental contact between the wires is more important than any other. Two wires must never be run through the same opening in a ceiling without the use of hard rubber separating-tubing slipped over each wire. The same is true of walls and partitions, where, if practicable, earthenware separating-tube should be used.

Are these Precautions Needful?—The questioner has only to read the rules laid down by the fire offices and the suggestions of the Institute of Electrical Engineers to find an answer. He must bear in mind that *although electricity is the safest illuminant ever used*, it consists of energy conveyed in wires, and that it will either manifest itself as light or as *heat.* If too much of it be forced through a thin conductor, that conductor will become hot, and it may become red-hot. If it can find a short path back to the mains *without passing through the lamps,* it will inevitably do so (as in two wires crossing). This will shortly—unless the insulating covering of the wires be very good—cause a contact and an electric arc, which may possibly give rise to fire. But when electricity is compared with gas, it is both easier to make it perfectly safe, and to provide *beforehand* for *leakage.* A gas-pipe may leak and suffocate every one in a house; an electric wire, if it leak, would heat up its fuse, and *get cut off* from the supply. There is no such possible precaution in the case of gas.

A great deal of nonsense has been spoken of the dangers of electricity. Although it has been in extensive use as an illuminating agent in this country for at least ten years, it is difficult to point to a single authentic instance of damage due to it. As used in houses, at a pressure of one or two hundred volts, it is perfectly harmless to the person. The conductors are so insulated and protected by cut-outs that any accident that might cause a fire is rendered impossible. Provided then a conductor of sufficient size, so that its sectional area is from one to one and a half square inch for a thousand lamps (or from 1000 to 1500 ampères), and suitable fuses inserted at the root of each branch, danger is entirely out of the question. But a good deal of discreditable work in the form of wiring has been done by unscrupulous contractors. Insufficient insulation has been put upon the wires. These have been carelessly run. They have been loaded with current (possibly from 2000 to 4000 ampères per square inch), so that they were always hot when at work. They have not been protected by fuses at all. And thus, through general ignorance, many installations have proved unsatisfactory, and broken down after a time, the users returning in disgust to gas or candles. Happily this state of things is passing away. Well insulated wires are being introduced, having ample carrying capacity, and their distribution is now better schemed. Fuses are being employed with many other precautions. There is one leading maxim for a contractor putting in electric light, and it is to avoid contracts that do not allow of the best class of material and labour being used throughout.

Tests during Wiring.—As suggested at p. 151, the general plan of the wiring must be taken upon paper, together with particular attention to such details as the positions of fuses and switches. During the progress of the wiring the leading hand should every day test each circuit as it progresses for continuity and crosses. He can find crosses or short circuits most easily by earthing all the distant ends of wires according to the directions given at p. 83. Continuity and freedom from short circuits having been ascertained, the final consideration in an installation of any considerable size is the resistance, both copper and insulation, of the circuits. Several hints as to these tests are given at p. 84, together with particulars of the instruments required. The tests should always be made from the dynamo, or from the point where the branch mains enter the house. In taking the copper resistance tests the ends of the far branch leads and returns must be twisted together, or connected with brass screw-junction pieces. All lamps must be removed. Every successive step in the testing must be made according to the plan of the wiring, which should be placed upon a wall near to the main switch-board. It is usual to take the insulation tests last. It may be of interest to state that a great deal of very good incandescent wiring has been done without taking either copper or insulation resistance tests. But in such cases the copper conductors have in every case been carefully selected to suit the distances at the outset. The insulation has been of the best, and the work in all its details carried out under the eye of the responsible electrician. Testing, at best, is but the detection of possible careless work or unforeseen accident. Continuity tests cannot well be dispensed with. Insulation tests are essential aboard ship, or in mills, and in any situation where there is danger from damp or contact with wires.

Prof. Jamieson's Rules for Insulation Resistance of Electric Light Circuit.—In a paper read before the Institute of Electrical Engineers,* by Professor A. Jamieson, F.R.S.E., he gives the following formula relating to the insulation resistance which should exist in the best kind of installations of the electric light :—

Let R_I = the total insulation resistance of the whole or any part of the lamp circuits, or of the generator, in ohms ;

K = a constant, $\cdot 1 \, \Omega = \frac{1}{10} \Omega = 100{,}000 \, \omega$ (100,000 ohms) found from actual tests of several well-erected installations ;

E = E.M.F. of dynamo or installation in volts ;

N_L = number of lamps (16-candle power) on each circuit or on the whole circuit, then

$$R_I = K \frac{E}{N_L}$$

The insulation resistance is therefore here taken to be *directly* proportional to the nominal E.M.F. of the dynamo, and *inversely* proportional to the number of 16-candle power lamps in circuit.

The Phœnix Fire Office rule puts the insulation resistance of different sized installations into tabular form, as follows :—

* Journal of the Institute, January, 1889.

184   WIRING FOR INCANDESCENT LAMPS.

Installations of	25 lights	. . . .	500,000 ohms.
,,	,, 50 ,,	. . . .	250,000 ,,
,,	,, 100 ,,	. . . .	125,000 ,,
,,	,, 500 ,,	. . . .	25,000 ,,
,,	,, 1000 ,,	. . . .	12,000 ,,

This applies to continuous currents having an electromotive force of 200 volts and under, and implies a test taken at one operation over the whole installation.

It is, of course, well known that tests of insulation are the exception, and, unfortunately, not the rule. It should be urged that insulation testing is quite as important as continuity testing, and certainly more important than conductor resistance testing, especially aboard ships. If insulation tests are to be neglected, the greatest precautions must be taken in the matter of selecting well-insulated wires, and in running them in the safest positions. See also "Rules of the Institute of Electrical Engineers," p. 212.

Estimation of the Electrical Power Required.

All electrical work, in wires and at lamps, represents the expenditure of mechanical energy. The mechanical units of measurement cannot, however, be employed in calculations of the electrical work. The electrical units employed by practical engineers in estimating electrical work are named as follow :—

Volt.—The accepted unit of measurement of electromotive force, or the potential difference between the poles of a machine (very generally regarded as, and styled, "pressure"). This unit bears a certain relationship or proportion to the absolute unit of pressure, the physical significance of which is fully explained in most text-books of electricity. *Voltmeters*, showing at a glance the voltage of any electric source, are

ESTIMATION OF THE ELECTRICAL POWER. 185

generally graduated by means of a standard galvanic cell, the electromotive force (as volts) of which is constant and well known. Either the ordinary telegraph Daniell cell, the volts of which are approximately 1·07, or Clark's standard mercury cell (volts 1·434) is used as a standard of comparison. The electromotive force of any electric source (dynamo, accumulator, &c.) is really a potential condition, and cannot correctly be regarded as similar to head of water or other mechanical pressure. It may be regarded as the state of strain existing between the terminals of the dynamo *tending to set up a current*. If the current be allowed to flow the strain is at once relieved, so that in measuring the potential difference between the terminals the voltmeter (although connected across them) is of so high a resistance as to prevent the setting up of a sensible current. The volt is generally symbolised in formulæ by the letter E (electromotive force). It bears a certain practical relationship to the other units of electricity spoken of below.

Ohm.—Electromotive force or pressure cannot exist unless there be a certain *Resistance* to the flow of electricity. Every conductor offers a certain resistance to flow. This resistance is measured in terms of the unit named after the famous enunciator of the law of electric circuit, Dr. Ohm. Its physical significance and derivation are explained in most of the text-books. The ohm is the resistance offered by a column of mercury 1 square millimetre in cross section and 106 centimetres long. 210 feet of No. 16 standard wire gauze copper wire, at a temperature of 60° Fahr., exhibits a resistance of one ohm, symbolised R (resistance).

Ampère.—The third factor dealt with by Ohm's law

is that of current, or flow. It may conveniently be expressed as the electric flow that would occur if a volt pressure were applied to an ohm of resistance. This current is called an Ampère after the celebrated French mathematician of that name. It is generally symbolised C (current).

The expression of these units brings us to the relationship they bear to each other. If, now, a volt of pressure be set up at the terminals of a dynamo, and a wire measuring an ohm be made to connect them together, the current flowing will be an ampère, as explained in other words above. The law is variously expressed

$$\frac{\text{Electromotive force in volts}}{\text{Resistance in ohms}} = \text{current in ampères}.$$

or the pressure divided by the resistance gives the current. The law may also be written—

$$R = \frac{E}{C}$$

or the resistance can be found by dividing the pressure by the current ; or it may be expressed—

$$E = C \times R.$$

In different words, the current is directly proportional to the electromotive force exerted in, and inversely proportional to the resistance of, the circuit.

Relation of these Units to the Mechanical Power.—To set up a current through a resistance, energy must be expended; this is called Power, or Work. The unit of electrical activity in a circuit, bearing a direct relationship to the work of the steam-engine, is called a *watt* (746 watts = 1 electrical horse-power). The watt is really a volt-ampère. As an engineer speaks

THE KILOWATT-HOUR. 187

of so many foot-pounds so does the electrician speak of volt-ampères, or watts. When it is required to measure the work done by a current in a wire or a lamp, it is necessary to ascertain the ampères of current flowing through it, and the volts of pressure impelling the current. The two numbers so found multiplied together gives us the activity or power in watts or volt-ampères.

For example, if it be required to determine the power expended in maintaining a certain number of lamps in a circuit, the voltmeter shows a pressure of 50 volts, and the ampèremeter a current of 15 ampères, multiplying these together gives the watts 750, which, divided by 746, the number of watts in an electrical horse-power, shows that the circuit is consuming energy of a trifle over 1 horse-power.

Again, the rule may be applied to a dynamo to ascertain its output. If the voltmeter shows 100 volts, and an ampèremeter 10 ampères of current flowing in the lamp circuit, the dynamo is yielding 1000 watts. This output is called, under the provisions of the Board of Trade regulation, a *kilowatt*, and if the power so expended be continued for an hour it constitutes a *kilowatt-hour*, which is the unit now used for electric lighting, in the same way as " 1000 cubic feet" is employed for gas lighting. A machine yielding a kilowatt would be known as "a one-unit dynamo." The machine is supposed to run at a suitable speed, and to maintain the current for long periods without heating. The performance of the dynamo under these conditions is called its capacity. Such a one-unit dynamo would light about 20 lamps, each taking 50 watts. As lamps are now made each would probably give a light of 20 candle-power, the watts per candle-power being 2·5. Such a machine would yield, according to the definition of the watt, 1·34 electrical horse-power.

The Electromotive Force (Pressure) Required.

For 50-*volt Lamps.*—The volts of pressure to be put upon the circuits will depend upon two conditions: (1), upon the voltage of the lamps, and, (2), the resistance of the circuits. In a small installation 55 volts should be ample when 50-volt lamps are used. This allows 5 volts for fall of potential due to resistances and for increased fall of potential due to increased current when all lamps are lighted, which is a large allowance. Five per cent. is usually considered a large allowance from dynamo to lamps and back. If the wiring be schemed according to the directions already given, it will cause a fall of pressure of about 2·5 volts for every 100 yards run. That is, if the most distant lamp be 50 yards away, 2·5 volts only will be lost in leads and returns.

For 100-*volt Lamps.*—These lamps are more economical to run than 50-volt lamps, but have not the same "life." The 100-volt lamp is very generally used in this country. An allowance of 100 volts, at least, is made for lamps, with the usual 5 per cent. additional for fall of pressure. The greater the number of lamps the greater the fall of pressure. This is due, as above explained, to the necessarily increased current. The fall of pressure of course is a characteristic of each installation, and cannot be exactly determined unless all the details of the wiring are known. Many of the first electricians scheme their wires on the basis for current of from 1000 to 1500 ampères per square inch of section, proceed to run them, and make a voltage allowance after taking the resistances—by methods already given—first, with all lamps "off" and parallel wires temporarily connected at their far ends; secondly, with all lamps connected and terminal wires disconnected.

The Current (Flow) Required.

In estimating for wiring and lamps it is necessary to consider that we are arranging for the continuous consumption of *power*. If we spend as little money as possible upon copper conductors—just enough to keep them from overheating—we are arranging for the *maximum of waste* of power. The conductors should always be as large as possible, or as convenience will allow. This will ensure the minimum of waste on the wires. The suggestion of Sir William Thomson that *conductors should present an effective conducting sectional area of a square inch for each* 1000 *ampères* of current carried, is only a suggestion made for the protection of buildings from fire; it does not imply that 1000 ampères per square inch is the best proportion. Many installations are running at 1500 and 2000 ampères per square inch of section of conductor. At the latter current the copper would become warm and would tend to soften the insulation. At both volumes of current great waste of power is incurred in the conductors. The user of the electric light would speedily find that 500 ampères per square inch of conductor was a more economical system of wiring than any larger proportion. But, as above stated, in house wiring conductors are short, and must, from considerations of bulk, be kept thin, so that for such work Sir William Thomson's rule is probably the best.

The current taken by lamps varies considerably for different lamps of the same nominal voltage and "watts per candle." The usual 50-volt lamps take approximately *an ampère each*. The ordinary 100-volt lamps of 20 candle power take approximately *half an ampère each*.

It will thus be discerned that in estimating a wiring system the voltage of the lamps to be used must be known.

But this is rather an unsatisfactory and rough method of arriving at the candle power and current. It will be seen from the table at p. 146, that the current and volts may vary considerably. But it is the custom to work the lamps at the highest practicable pressure, so that 100 volts is quite commonly put upon 16 c. p. lamps, and 120 volts upon 20 c. p. lamps; hence the "watts per candle" (p. 170) is as low as possible. The practice of the Edison-Swan Company is to indicate all lamps taking more than ·9 ampère at 4 watts per candle, and all lamps taking less than this amount at 3·5 watts per candle.

Roughly, an electrical engineer allows half an ampère per 100-volt lamp and one ampère per 50-volt lamp, in estimating his dynamo power.

According to this approximation the volts and ampères per 100 lamps of the 50-volt class will be, allowing for "fall," 55 volts and 100 ampères. For 100-volt lamps 105 volts and 50 ampères.

Methods of Jointing the Conductors.

Materials required.—A jointing tool and material case, containing suitable receptacles for all the usual

JOINTING OF CONDUCTORS. 191

tools and material; or a leather satchel, as used for linesmen's tools.

1 small bench vice, 1 hand vice.
1 insulation knife, 1 scissors.
1 flat file, medium cut.
1 pair flat-nose cutting pliers, 1 ditto plain, 1 ditto round nose.
1 lb. tinned wire, fine, for binding joints.
1 soldering iron.
1 portable soldering furnace, 1 appliance (Bunsen burner) for heating wire by gas, 1 spirit jointing-lamp.
$\frac{1}{2}$ lb. solder.
$\frac{1}{4}$ lb. resin, in a box.
1 small bottle Baker's soldering fluid, or solution of zinc chloride.
3 sheets "F. F." emery cloth.
1 tin of india-rubber solution, 1 do. Chatterton's compound.
1 lb. each of $\frac{1}{2}$-in. and 1-in. thin sheet wrapping india-rubber and india-rubber tissue.
1 lb. felt tape (compounded) for wrapping.
1 bottle strong, thick shellac varnish, 2 brushes.
1 ball spirit-lamp cotton, 1 tin best wood naphtha.

Instruments.—If the wiresman is also intrusted with the testing of his circuit (for continuity), he will require a linesman's galvanometer or detector, and a small dry battery of about six cells. Both the Leclanché and the chloride of silver cells are used for this purpose. These are usually combined in one case, with the necessary connecting wires.

Method of making a Common Joint in Gutta-percha-covered Wire.—Cut away the insulating material from both wires for about 2 inches. Do not notch the wire in doing this. Scrape the ends quite clean.

Place one conductor across the other near to the insulation, and grip fast with pliers or hand-vice according to size of wire. Twist the conductor ends over each other alternately until a neat, close spiral is obtained, at least $1\frac{1}{2}$ in. in length. Clip off the remaining copper ends, and trim smooth with the file. Again clean the joint by scraping. Apply a *very little* soldering fluid (or preferably, resin—see "Soldering"), and tin the joint with the iron. Wipe carefully, especially if fluid has been used, and it has not been all "burnt out" in the tinning. Proceed to stretch down the insulation from either side, over the joint. Keep the gutta-percha warm over the lamp while doing this. Tool the gutta-percha together with a warm iron where it meets, and allow it to set before finishing the joint. Put on a coat of Chatterton's compound in the middle of the joint, and allow to set. Take a strip of thin gutta-percha sheet several inches in length and an inch wide. Warm this up and attach one end to the joint. Keeping the rubber soft wrap it round the joint—it will form an enlargement. Before it cools work the wrapping in both directions with thumbs and fingers until it extends completely over the joint—it should be slightly thicker than the ordinary size of the insulated wire when done. Tool it smooth with a warm iron, leaving it smooth and compact. The joint should be capable of withstanding immersion in water even longer than the general insulation. It is essential that the hands and materials be clean.

Joint in an ordinary Taped Lead.—Unwind the insulation 4 inches from each end and strip off 2 inches of the interior gutta-percha. Scrape the metal clean. Splice or scarp the ends for $1\frac{1}{2}$ inches with

JOINTING OF THE LEADS.

the file, so that when placed together they do not form a joint larger than the wire. Hold one end in the vice, place the other upon it. Touch with resin or soldering fluid and carefully solder together. File round, and bind upon the joint a close spiral of tinned wrapping wire. Solder all together. The covering will depend upon the insulation of the wire. Heat a long strip of gutta-percha and wrap it quickly around the warm joint. Work it lengthways until it combines with the insulation. Tool it down with a warm iron. Cover with a close winding of india-rubber coated tape, with coatings of india-rubber solution between. The exterior covering is generally compound-coated tape, carefully wound in several layers, combined with the tape unwound from the wire, with shellac varnish between each layer. Over all a coating of varnish. The joint must withstand continuous immersion in water.

A T-joint or Branch from Main Lead.—Strip, by unwinding, the insulation of the main lead for about 2 inches. Clean the copper. Strip the branch extremity for 4 inches, and clean it by scraping. Place the branch *across* the main at right angles. The branch should touch close up to its insulation, and it should cross the main to the extreme left of the bared portion. Hold in position with the pliers. Proceed to wind the branch around the main in a tight spiral. Apply resin or solution and solder together carefully. Wipe clean. Apply coating of shellac varnish. Wind cotton insulation from both conductors alternately around the main. Shellac and varnish over all. Wind on a strip of gutta-percha sheet while soft, and draw it well over the joint. Tool it down. Cover with several close windings of india-rubber-coated tape, with coatings of solution between. Allow the solution to dry before proceeding. Over all wind two coverings of felt tape, with shellac varnish. The joint must be perfectly clean and smooth, and only slightly thicker than the main conductor. If the branch is liable to longitudinal strain, bring its end back and wind it several times around itself while winding bare upon the main. If the main is stranded, solder the strand together before winding in the branch.

Joint in a stranded Conductor.—Clear the insulation from either end for 3 inches. Separate the central strands and cut out both the central wires. Twist each pair of wires together separately. Twist all together and solder carefully. Make the insulation joint as before.

General Suggestions.—The operation of making a good joint calls for considerable practice. Carefully stripping the insulation, so that it may serve to lap the joint, is an important point. Cleaning and tinning must be thoroughly done. Connecting or splicing must be neat and strong. Wire wrapping must be close and neat. Tinning over all must be effective. The subsequent operation of insulating the joint is of the greatest importance in wiring. The wiresman must not forget that he is making *two* joints at one time—a metallic conducting joint and an *insulation joint.* The insulation joint must be quite as effective as the metallic joint. This is chiefly produced by the skilful use of soft gutta-percha, made to unite with the gutta-percha of the wire, or with tape and varnish, forming when dry a solid coating of insulation. Unless the joints in a lead be specially made to resist strain such a lead must never be subjected to tension after being jointed. The main outlook in jointing is to produce continuity of conductor and continuity of insulation *at least* as good as that of the general run of the lead. Most skilful hands produce joints much more conductive and better insulated than the general run of the wire.

Soldering and Tinning.—A much-contested point amongst electrical engineers is whether resin or soldering fluid should be used as a flux. We think an answer may be readily found in the following :—If the workman is not thoroughly acquainted with the use of resin, in tinning copper, let him use fluid. A resin joint, unless very well made, is very deceptive, and may appear to be sound, when a coating of resin may exist between the ends to be united. If resin is properly and sparingly used, it no doubt makes the best joint for keeping, or permanency. The objection to chloride of zinc solution is that it is sometimes left upon the joints and may set up electrolytic action when current is on, speedily destroying the joint. If chloride of zinc be left on a joint that joint will never become dry. This salt is one of those that absorb moisture from the atmosphere. If left upon iron, or, indeed, almost any metal, it tends to set up oxidation—in the case of iron very rapid. *But there is no flux so certain in its action as soldering fluid.* There is a variety of it believed to be free from most of the objections to chlorine of zinc, known as Baker's tinning fluid. This kind is used exclusively by the Post Office electricians. If joints made with fluid can be washed and dried afterwards they will be quite safe; such a joint is never deceptive. The fluid "flux" will make a good joint on a surface so dirty that resin would never permit the iron to tin. The main provision in soldering joints is cleanliness—pure, bright copper, untouched by hand.

All joints should be soldered.

Electric Lamps on Gas Fittings.—The failures that have resulted from making gas fittings act as a return wire, in some installations of the light, should be sufficient to dissuade any further experiments of the kind. *Gas fittings are not necessarily conductive* throughout—joints are generally faulty points.

But such a fitting as a chandelier may generally be made to serve as a "return," within itself only. That is, the two wires are brought to it, and one of them soldered to the ceiling fixture of the pendant, the other being taken downward to the lamps. Wire for this purpose must be very heavily insulated, and it must not be led or drawn over sharp angles of the metal, otherwise it may set up short-circuiting. When gas fixtures are to carry electric lamps only, the gas pipes should be disconnected from them, otherwise short circuits are apt to be got throughout the system.

If both gas and electricity are to be used, the fixtures should not, except within themselves, be used for return, and in any case it will be essential to observe that the *negative* wire (if that wire be regarded as the "return") is in every case soldered to the fixture, and not any wire at random, whether negative or positive. *Insulated joints* for insertion in gas systems are coming into use for cutting each fixture off electrically from the main system of pipes. The gas supply is kept up through a tube of insulating material.

All electric light wires must be kept a safe distance from gas or water pipes.

A fusible plug must be placed in the ceiling plate, above each pendant carrying more than one light.

It may be broadly stated that all electric light fixtures must be insulated from any metallic support they may have.

In the ordinary wiring of a chandelier the lead and branches are simply led along the course of the arms, concealed as much as possible. If the wires can be completely concealed, that is done by inserting them beneath the ornamental coverings or shells. One wire only is taken to each lamp. The other contact is got by soldering the negative terminal of the lamp, or its wire attachment, to the body of the chandelier. Thus, the pair of wires led to the ceiling opening over a chandelier will form a pair of "mains," and the branch wires the "branches."

In wiring gas fixtures in which there are joints the connection is kept up across the joints by short spirals of insulated wire, flexible enough to move with the joints without danger of breakage.

Electric light fixtures, intended for that purpose alone, are generally already wired by the makers, and are specially adapted, so that their fitting is a comparatively simple matter. There is usually provision left in the ceiling rosette, or wall-plate, to allow of the insertion of a fusible plug there—a precaution which is generally observed in the case of electroliers with several lamps attached.

The fitting of the incandescence bulbs themselves to the gas fixtures is greatly facilitated by the use of "*adapters*," consisting of screwed nozzles fitted to the lamp, and with gas thread to take the place of the ordinary burners, usually $\frac{3}{8}$ in. gas thread. But it is rather unusual and unnecessary to attach the glow lamps so that they take a vertical position, as in the case of gas burners. It is generally more effective to arrange them pendant from the chandelier. This is easily effected by the use of "*pendant arms*" with screwed nipples, to be fixed in the place of the gas burners. The lamps can, by these means, be arranged either pendant or at an angle of 45° with the vertical —a favourite and effective position for an incandescent lamp.

CHAPTER VI.

INCANDESCENT LIGHTING OF SHIPS.

THE rapid adoption of electric light aboard steamships has caused a considerable demand for information on the subject. We propose therefore to offer a few hints and suggestions bearing upon the condition of things suitable for use at sea. But the subject is so wide that ship lighting might well require a treatise of considerable size to fully do it justice. Premising, however, that the reader seeking this special information is already acquainted with what we have written on house lighting, it may be practicable to embrace the chief points of interest within even the restricted space at our disposal.

Dynamos.—For use aboard ship there can be no doubt that slow speed dynamos give the least trouble. The importance of providing a ship with electric machinery that calls for little attention can only be appreciated by those who have been at sea, where the electric lighting has usually to be looked after by the engineers in turn as they come on watch. The mechanical engineers, although perfectly competent in their profession, are not to be expected to pose as experts in the handling of dynamos or faulty circuits. Hence, apart from mere mechanical attention, the ship dynamo should not need any kind of supervision. This condition is perhaps better filled by a compound-

(series and shunt) wound dynamo, running at a slow speed, than by any other kind of machine. It must give a self-regulating current and pressure suited to the lamps. The compound-wound dynamo, if well designed, can be made to regulate so closely that if half the lamps be suddenly turned off or on scarcely a flicker will be observed in the remaining lamps. A ship dynamo should be self-regulating down to at most 10 per cent. of the lamps.

If 100-volt lamps be used, the dynamo is generally selected to give a pressure of 110 volts at least, with a minimum of half an ampère of current for each lamp. Thus, if 1000 hundred-volt lamps are to be run, the dynamo must give at normal speed 500 ampères at 110 volts.

If 50-volt lamps be used, as is generally the case aboard ship, the dynamo must give a current of 1000 ampères at 55 to 60 volts. These figures are approximate only, depending upon the "watts per candle" of the lamps to be used; upon the size of the leading wires and the insulation employed. The lower pressure and larger leading wires are doubtless most suitable for ships; but the expense of running the light is somewhat greater than when high voltage lamps are used. The volts of the dynamo should in no case be less than 50, on account of the necessity for the use of arc lamps aboard ship, for canal navigation and unloading or loading at night.

Driving.—A good deal of controversy has taken place as to the comparative advantages of belt and direct driving. There can, we think, be little doubt that direct driving—attaching the dynamo direct to the engine shaft—is the most generally applicable plan. Several special engines have been designed for this purpose. A special separate engine, nicely self-regulating, is an absolute necessity. The main engines of a steamship in rough weather run at all speeds, and cannot be utilised for driving a dynamo if the machine is used for lighting lamps direct. Main-engine driving has, however, been tried where the lamps have been run off an accumulator. But if it comes to a question of special engine *versus* accumulator, the engine has decidedly the best of the case. A large battery of accumulators is scarcely suitable aboard a ship, unless the vessel carry a qualified electrician to run the plant.

While speaking of accumulators it may be mentioned that a small battery of them is extremely useful for keeping the "all night" lights going, after the dynamo has been stopped. Such a battery is usually charged by running the dynamo upon it during the day. But for ship work a battery of 26 cells appears quite sufficient. Such an installation can only be run satisfactorily when tended by a man acquainted with both dynamos and accumulators. For further information as to the running of accumulators the reader is referred to Chapter II., p. 42.

Disturbance of the Compasses in Ship Lighting.

The "Nautical Magazine" for December, 1885, contains a contribution from Mr. William Bottomley on the subject of the probable interference with the compasses by the currents used for the electric light. The following example of a supposed case showing the amount of error which may be produced on the compass unless precautions are taken to guard against it is there given:—

Suppose a main lead from the engine room to the fore part of the ship to light up 100 lamps is brought along the centre of the ship. It may be at a distance of 10 metres or 33 feet from the standard compass, and will run almost underneath it. If we suppose that each lamp takes one ampère of current, there will be a current of 100 ampères altogether in this lead.* Now the effect on the compass at a distance D in centimetres is given by the formula

$$F = \frac{2 \times \frac{1}{10} C}{H D},$$

where C is the current in ampères and H is the horizontal magnetic force. In this case we have C = 100 ampères and D = 1000 centimetres. Therefore

$$F = \frac{20}{1000 \, H} = \frac{0·02}{H}.$$

At Glasgow the horizontal force may be taken as 0·15 in C.G.S. units. Therefore the effect upon the compass will be $\frac{·02}{·15} = \frac{1}{7·5}$. This will be expressed in degrees by multiplying by 57·3 the number of degrees in the radian, or angle subtended at the centre of a circle, by an arc equal in length to the radius. Therefore the amount of error produced by such a current on the compas will be

$$\frac{57·3}{7·5} = 7·6 \text{ degrees.}$$

The foregoing refers to a single wire and a continuous current machine, but if an alternate current machine is employed no effect will be produced on the compass, even when the ship's side is used for the

* It is unusual, however, to put more than 50 lamps upon a single lead.

return. When a continuous current machine is used the danger of producing an error on the compass can be avoided by using two wires close to one another, but these wires should be well insulated from the ship's side. If in any way one of the wires is brought in contact with the iron of the ship, there may be no change observable in the lighting, but the current may produce as much error on the compass as it would if there was only a single wire.

The following points should therefore be attended to in cases of lighting ships by electricity :—

When a continuous current machine is used the circuit should consist of leading and return wires, as in house lighting, with this difference, that the wires should be kept close together wherever practicable.

Insulation resistance should be tested periodically to ascertain if there be any leakage to the ironwork of the ship.

These precautions are recommended because in ship lighting, as commonly carried out, only a leading wire is used, the "return" being effected through the shell of the ship itself, $e.g.$, no negative wire is used.

In the case of an alternate current machine a single wire may be employed, and the iron of the ship used to complete the circuit without producing any effect upon the compass.

Error not readily Detected.—The question assumes greater importance when we consider that the error of the compass due to the electric lighting is not liable to come under the notice of the officers of the vessel. The routine is to determine the error (the usual working error) of the compass during the day, while the electric light is not employed. The error may thus be determined as usual every day, and the course of the ship set by these determinations; but when the electric light is turned on, the course of the vessel may be changed, and before the light is turned off she may be several degrees out of her path.

Mr. Bottomley refers to the case of a dynamo being placed near to an iron bulkhead, the upper end of which happens to be near to the compass. It is assumed that the bulkhead may become so strongly magnetised by the field magnet of the dynamo that a considerable error may be produced on the compass.

In the discussion that followed a paper read by Mr. Bottomley at the Society of Arts,* in which several authorities on the subject took part, including Captain Creak, of the Admiralty Compass Department, the case of three ships lighted on the single-wire system with continuous currents was cited, in which distinct error had been observed upon the compasses when the lights were turned on.

Sir William Thomson † mentions cases of large passenger ships lighted by continuous current on the single-wire system, in which as much as 4° and 5° of error on the compasses had been produced by the electric lighting. In the latest of these cases an error of 4° on the north course was found when the light was turned on. The light was put on and off several times with the ship's head north, and every time the same error was produced.

Mr. Alexander Siemens, another well-known authority on this subject, points out that in calculations that have been made to show the effect upon the compass of the electric-lighting current, the screening effect of the iron decks had been neglected. He is

* "Journal of the Society of Arts," Feb. 5, 1886.
† See paper on the subject read before the Institute of Electrical Engineers, May, 1889.

of opinion, judging from practical experience, that two wires are unnecessary, and that the disturbance of compasses is only brought about by running single leads close to them, or situating the dynamo in immediate proximity.

Captain Creak instanced the case of a war ship, in which the direct compass disturbance due to the generating machine was appreciable at a distance of 55 ft., and across iron bulkheads, and that it was perceptible also in ships of the Royal Navy lighted on the two-wire system. In the case of H.M.S. *Northampton* there was considerable trouble due to "magnetic leakage" from three dynamos placed in such a position that the external field produced was directed towards the compass. The error on the compass was 11° when all the machines were running, and three correction tables were necessary.

Test for Compass Disturbance due to the Currents.— It is now usual, since attention has been called to the subject, to apply a test before the ship leaves dock to have her compasses adjusted. This is simply effected by putting on and off the lights, and observing the compass. This is done as the needle stands, and also after it has been artificially deflected from its position to the extent of 45° on either side of zero by means of a small permanent magnet suspended above the glass case.

It is clear, from the extensive experience that has been gained on the subject, that by the use of dynamos diffusing little magnetic leakage—possessing a compact magnetic field directed only upon the armature—as now constructed the faults due to this factor can be overcome. The dynamo must of course be kept as far as practicable from the compass.

There can be no doubt also that hundreds of ships have been supplied with electric light on the single-wire system, in which the error on the compasses is very slight, if at all appreciable.

This is usually effected by keeping the leading wire as near to the ship's side as possible, and by observing that some iron screen, as the upper deck, interposes itself between current and compass.

Ship Wiring.

So much has already been said with regard to house wiring that only the main peculiarities of the methods observed in ships need be noticed.

Single-wire Work.—Since there is no return wire, and since, by connecting the negative poles of the lamps to the iron work of the ship, the negative is practically earthed, *very good insulation* must be used. The wire employed must first have a conductivity at least as high as 95 per cent. It must be insulated in a very thorough manner with pure india-rubber, combined with cotton insulation or other substance, and completely vulcanised, as spoken of at p. 150. The exterior protection must be strong, and adapted to withstand the roughest handling. The rules given at p. 86 will apply to the covering of the cable, and at p. 146 to the selection of a conductor of sufficient carrying capacity. In ship wiring, when the wires can scarcely be kept free from damp, the insulation must be especially effective.

The general opinion is in favour of simple cleat wiring, without casings, when it can be employed, below decks. The cleats used are single-groove cleats.

Connection with the dynamo is got by bringing a thick piece of cable from its negative terminal and making a good connection to a large copper or gun-metal stud screwed into the iron work of a bulkhead, or any main piece of the ship's shell. The leading cable's insulation covering is kept away from actual contact with the iron work. When it is run along the ship's side or under an iron deck "runners" of wood are interposed. These runners should be varnished or soaked in melted paraffin wax, to prevent the absorption of moisture. It is usual to run the lead under the main deck, and to take from it branch wires to the lamps to be fed. In the case of a single lamp, situated away from other lamps, its negative or *return* would be made by attaching the terminal by means of a short piece of copper wire to a brass stud screwed into the nearest iron work. In the case of groups of lamps one such stud is made to answer for several lights. These are known as " return studs."

Joints in the leading cable are usually not only made with extra care, but are afterwards protected by cast-iron joint boxes, and the same boxes are generally fixed over connections to branches, at which point also a fuse is of course inserted.

Precautions against fire are taken by the use of double-pole fuses at the dynamo, and the insertion of one at every branch root, and frequently a fuse is inserted for single lamps. A fuse is of course put to every cluster of lamps, as in house wiring.

The wiring of ships is generally carried out on the parallel system (p. 125), with only one lamp between lead and ship. The tension employed, as already stated, is not often over 50 or 60 volts.

The two-wire system is not so extensively used as

it should be. When it is installed the insulation of the wire need not be so heavy as is required in the single-wire system, and with a view to the protection of the compasses the lead and return should, when practicable, be run within a few inches of each other.

Saloon and cabin wiring must be done under casings or mouldings, as explained at p. 177. There is one point in wiring a ship that should receive attention. Insulated wire does not last in use indefinitely. It may have to be renewed every few years. For this and other reasons the position of the wires should be such that they may be accessible for purposes of repair or renewal.

Ship fittings are a class of themselves. They are usually of a very solid make. The bulkhead and engine-room and passage-way lamps are placed behind glass screens, and protected by iron gratings. Side lights are also coming into use, fed from a branch taken from the nearest main. Each cabin lamp has its own switch. The saloon lights are controlled by the attendants by means of a main switch.

The main switch-board is generally placed in the dynamo room, which is usually the engine room, and is fitted to control (1) the " cabin's circuit," (2) the " saloon's circuit," (3) the " officers and men's circuit," separately, as may be convenient. Thus there are frequently a number of separate circuits run, so that any one section of lamps may be controlled separately. It is rarely that more than 50 lamps are placed in a single circuit.

Compass Electric Lamps.—These have been tried, but not with much success. Major Cardew stated[*] that he had carefully twisted the two leading wires to-

[*] Meeting of the Institute of Electrical Engineers, May 23, 1889.

ELECTRIC LIGHT PROJECTOR.

gether to eliminate the effect of their induction upon the compass, but that the current in the filament of the lamp itself affected the compass, and introduced an error. Better results have followed the use of metallic induction screens for obviating this inductive effect, and there can be no doubt that binnacle lamps lighted by the current will soon become the best possible source of light for a compass in rough weather.

Suez Canal Projector. — Owing to the successful use of the electric arc light the traffic has been carried on through the Suez Canal by night for some years past.

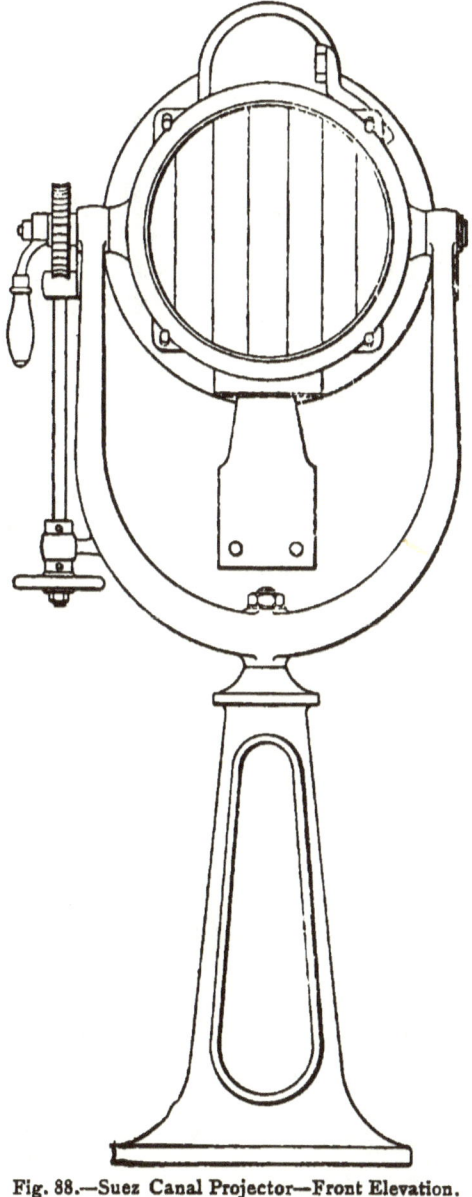

Fig. 88.—Suez Canal Projector—Front Elevation.

The arc lamp is placed within a projector usually

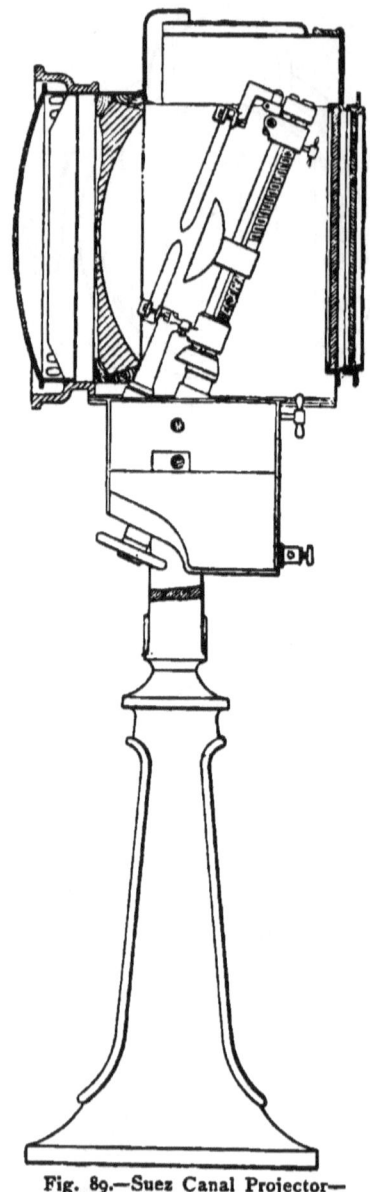

Fig. 89.—Suez Canal Projector—Section.

arranged as in Figs. 88 and 89, which represents the form fitted to ships by Messrs. Paterson & Cooper. The projecting portion, shown in front elevation in Fig. 88, and in section in Fig. 89, is controlled by the hand wheel and worm-gearing shown. The front of the projector is fitted with the usual optical appliance for concentrating the beam in a given path. These projectors are generally slung over the bow of the ship, at such a height above the water as will give the best effect. The projector is generally placed in a cage, and is kept in adjustment by an attendant who occupies the back of the cage. The maximum current put upon the projector is 60 ampères at a pressure of 50 volts, and it is now common to run it off the dynamo employed for the general lighting of the ship. A "choking coil" is usually placed in series with the arc lamp, absorbing about 15 volts.

Electric projectors and apparatus are frequently hired by vessels passing through the Canal, when they do not carry a dynamo, or when the lighting plant aboard is working up to its full power on the general lighting of the ship. The night navigation of the Canal is very greatly facilitated by the free use of arc lamps placed along its sides for considerable distances. If this system were extended it is probable that the use of projectors aboard the vessels would be unnecessary.

CHAPTER VII.

MISCELLANEOUS INFORMATION.

Abstract from the " Rules " recommended by the Institute of Electrical Engineers.—The committee appointed by the Institute, consisting of gentlemen well known in electrical circles, make, amongst others, the following suggestions, chiefly bearing upon the fire risks of electric lighting :—

Conductors.

Carrying Capacity.—Conductors must have a sectional area and conductivity so proportioned to the work they have to do that if double the current proposed is sent through them the temperature of such conductors shall not exceed 150° Fahr.

Accessibility. — The conductors or their casings should be placed in sight if possible, and they should always be as accessible as circumstances will permit.

Insulation.—Within buildings they should always be insulated, and this rule applies equally to all conductors and parts of fittings which may have to be handled.

Highest Permissible Temperature.—Whatever insulating material is employed it should not soften until a temperature of 170° Fahr. has been reached, and in all cases the material must be damp proof.*

Casings.— When conductors pass through roofs, floors, walls, or partitions, and when they cross, or are liable to touch metallic substances, such as bell wires, iron girders, or pipes, they should be thoroughly protected by suitable additional covering; and when they are liable to abrasion from any cause, or to the depredations of rats or mice, they should be encased in some suitable hard material.

Portable Lamps.—In the case of portable lamps and fittings with which flexible leads are used special precautions must be taken.

Distance Apart.—Conductors should be kept as far apart as circumstances will permit, the spacing between them being governed by their potential differences.

Inflammable Structures.—When conductors in very inflammable structures precautions should be taken to isolate them therefrom.

Metallic Protection.— Conductors which are protected on the outside by lead or metallic armour of any kind require the greatest care in fixing, on account of the large conducting surface, which would become connected to the core in the event of metallic contact between them.

Dangers from Operations through Walls.—In cases where conductors pass into a building—from one building to another, or from one room to another— precautions should be taken to prevent the possibility

---

* It may be noted, in reference to this rule, that, as gutta-percha softens at 115° F., becomes plastic at 120° F. and melts at 212 F., it is practically excluded as an insulator of wires when unsupported by other, and infusible, insulating coatings around the conductor. Vulcanised india-rubber withstands much higher temperatures.

of fire or water passing along the course of the conductors.

Joints.—All joints must be mechanically and electrically perfect, to prevent heat being generated at these points. When soldering fluids are used in making joints the latter should be carefully washed and dried before insulation is applied.

Gas and Water Pipes.—Under all circumstances complete metallic circuits must be employed. Gas and water pipes must never form part of the circuit, as their joints are rarely electrically good, and therefore become a source of danger.

Overhead Conductors.—Overhead conductors, whether passing over or attached to buildings, must be insulated at their points of support. Precautions must be taken to obviate all risk of short-circuiting when they are likely to touch a building or other overhead conductors and wires, either by their own fall or by being fallen upon by other conductors.

Lightning Protectors.—In the case of overhead wires every main should have a lightning protector at each point when it enters or branches into a building.

Metal Fastenings.—Metallic fastenings for fixing conductors should be avoided; but, when unavoidable, some additional covering should protect the conductor from mechanical injury at such fixing points.

Insulation Resistance.—The insulation of a system of distribution should be such that the greatest leakage from any conductor to earth (and, in case of parallel working from one conductor to the other, when all branches are switched on, but the lamps, motors, etc., removed) does not exceed *one five* thousandth part ($\frac{1}{5000}$) *of the total current* intended for the supply of the said lamps, motors, etc. ; the test being made at the usual working electromotive force.

Switches.

The main switches of a building should be placed as near as possible to the point of entrance of the conductors, or to the generators of the current. Switches should be provided in both leads.

Bases.—Switches, commutators, resistances, bare connections, lamps, &c., must be mounted on incombustible bases. Cut-outs mounted on bases of wood rendered uninflammable are admissible. Vulcanite bases are undesirable in damp situations.

Cut-outs.

All circuits should be protected with cut-outs; and all leads from the mains, or small conductors from larger ones, must be fitted with cut-outs at their branching points.

When fusible cut-outs are used the section should be so situated within its frame that the fused metal cannot fall where it may cause a short circuit or an ignition.

For all main conductors a cut-out should be provided for both the "flow" and the return (+ and — leading wires); and the two fusible sections must not be in the same compartment.

Arc Lamps.

These must be guarded by lanterns or netted globes, so as to prevent danger from ascending sparks, and from falling glass and incandescent pieces of carbon.

Transformers.

When these are used to transform either direct or alternating currents of high electromotive force—that is, from or to an electromotive force of, say, 2000 volts—they, together with their switches and cut-outs, must be placed in a fire and moisture proof structure, preferably outside the building for which they are required. No part of such apparatus should be accessible except to the person in charge of their maintenance.

Distance between + and − Terminals.—The positive and negative terminals connected to such conductors should not be permitted to be nearer each other than 12 inches.

Heating.—Transformers which, under normal conditions of lead, heat above 150° Fahr., should not be permitted to remain in use.

Danger from Internal Contact.—Transformers should be so constructed that under no circumstances whatever should a contact between the primary and secondary coils lead a high E.M.F. into the building.

These amended rules (certain paragraphs not pertinent to our subject, or already abundantly treated, have been left out) were issued by the Institute of Electrical Engineers, April, 1888.

INDEX.

ABSORPTION of light by glass envelopes, 167
Accumulator attendants, hints to, 42
  insulation of, 42
  position for, 42
  starting of, 43
  charging of, 43
  electrolyte for, 44
  tests for full charge of, 45
  dynamo for charging, 45
  working hints regarding, 46
  rate of discharge, 46
  sulphating of, 47
  faulty cell in, 47
  short circuiting of, 48
  cut-out for, 49
  excess indicator for, 49
  automatic switch for, 49
  leads from, to dynamo, 50
  reserve cells of, 51
Accumulators and dynamos in parallel, 56
  voltmeters for, 59
  contact maker for, 61
  hydrometers for, 61
  switches for, 156
  in ship lighting, 201
Adapters for gas fixtures, 197
Adjustment of Brush system lamp, 97
Alternating dynamos in parallel, 54

Alternators in parallel, test lamps for, 56
Alternating current circuits, arc lamps in, 100
Alternating arc, singing of, 118
  versus continuous currents, 137
Ampère meters for station work, 67
  unit of current, 185
Arc light wiring and fitting, 88
  lamps in parallel, 91
  focussing, 91
  distance between carbons in, 91
  series running of, 92
  ampères required for, 92
Arcs in series, regulation of, 92
Arc lamps, volts required for, 92
  trimming of, 93
  Brockie Pell, 94
  rules for adjusting, 94
  Brush, adjustment of, 97
  in alternating current circuits, 100
  sizes of carbons for, 101
  n series with incandescent lamps, 101
Arc lighting circuits, 102
  cables, size of, 104
  leads, table of sizes of, 105
  naked leads in, 109
  circuits in mills, 111

INDEX.

Arc light projector for ships' use, 209
Arc circuit, ground leakage in, 104
Area lighted, 167
Armature, overheated, 29
   repairs, 39
   loose binding of, 39
   broken wire in, 40
   wire splice in, 41
Arms, pendant for incandescent lamps, 198
Arrangements, switching, 151
Arrester, lightning, 106
Artificial resistance for dynamos, 53
Asbestos insulation of commutator, 18
Attachment for portable lamps, 173
Attendants of dynamos, hints to, 26
Automatic regulation of dynamos, 6
Automatic governors, attention to, 31
   switch for accumulator, 49
Ayrton & Perry's spring voltmeter, 66

BALANCING resistances, 73
   Bank of lamps as a resistance, 53
Bead hydrometers, 61
Bearings, hot, 28
Belting and speeding dynamo, 14
   lacing of, 15
   surface and power, ratio of, 15
Best position for accumulators, 42
Binding of armature, loose, 39
Black insulation, significance of, 175
Boards, fuse, 166
Branch line and lamp switches, 157
Brockie-Pell arc lamp, action of, 94
Broken conductor, localising, 35
Brush dynamo regulator, 7
Brushes, point of least sparking for, 11
   lead of, 11

Brushes, for dynamo, 16
   pressure on, 16
   material of, 16
   periodic vibration of, 17
   burnishing of, 28
Brush system lamp, adjustment of lamp, 97
Bulbs, blackening of lamp, 169
Burned spot of commutator, causes of, 17
Burnt-out coils, 36
Butt joint in belt, 15

CABLES, mechanical protection of, 103
   insulation of, 103
   and wires, difference between, 103
   for arc lighting, 104
   jointing of, 190
Calibrating voltmeter, 63
Capacity of switch, 160
Carbons, distance between in arc, 91
   for arc lamps, size of, 101
Cardew voltmeter, 57
Case and moulding wiring, 177
Central station work, 1
   time curve in, 27
   time and current curves, 56
Charging accumulator, 43
   dynamo for, 45
"Choking" coils, 115
Circuits, arc lighting, 102
   planning of, 140
   closed loop, 141
   size of wire for, 143
   methods of running, 175
   Ohm's law of the, 186
Cleats used in wiring, 176
Cleat wiring, 176
   crossing, 176
Closed-loop circuits, 141
Combined switch and cut-out, 159

INDEX. 219

Commutator brushes, 16
  material of, 16
  treatment of, 17
  lubrication of, 17
  roughness of, 19
  returning, 19
  new, fitting of, 20
  sparking at, 37
Compass disturbance in ship lighting, 201
  estimation of, 202
  tests for, 205
Compass electric lamps, 208
Compound-winding of dynamos, 5
Conductor resistance, to take, 78
  stranded joint in, 194
  joints in, 190
Conductivity test, 84
  of wires, tests for, 149
Conduit tube for piercing wall, 111
Connections of the dynamo, 23
Constant current dynamo, 4
  position of the neutral point, 12
Contact maker for accumulator, 61
Continuous v. alternating currents, 137
Converters or transformers, 112
Coil in lamp, differential or shunt, 90
Coils, burnt-out, 36
Copper and alloy fuses, melting points of, 163
Crossing cleats, 176
Current, direction of, 25
  unit of, 185
  required for given lamps, 189
  absorbed by incandescent lamps, 170
Currents, alternating v. continuous, 137
Cut-out for accumulator, 49
  and switch combined, 159
  and fuses, main, 160
Curve, time and current, 56

DAMAGE by lightning, 106
  Damp transformer, remedy for, 119
Dangers due to defective wiring, 181
Differential shunt regulating coil in lamp, 90
Dioptric shades for lamps, Trotter's, 173
Direction of current from dynamo, 25
Description of the incandescent lamp, 168
Distributing box system, 141
Disturbance of compasses in ship lighting, 201
Double break switches, 152
  pole switches, 153
Driving of dynamo with rope, 15
  aboard ship, 200
Dynamo, hand regulation of, 6
  brushes, neutral point for, 11
  lead of, 11
  management of, 12
  foundations for, 12
  insulation of, 12
  erection of, 13
  speeding and belting, 14
  rope driving of, 15
  belting and power, 15
  belt lacing of, 15
  brushes of, 16
  pressure of, 16
  material of, 16
  brush, periodic vibration of, 17
  commutator, burnt spot on, 17
  treatment of, 17
  insulation of, 18
  connections of the, 23
  magnet, direction of current in, 24
  mechanical test of, 25
  attendants, notes for, 26
  heat and attrition of, 28
  bearings, heated, 28

Dynamo, overloading of, 29
  working, hints respecting, 30
  faults in, localising, 32
  earth fault at, 33
  periodic faults of, 33
  tests for leakage in, 33
    for internal fault in, 35
  intermittent faults in, 36
  coils burnt, 36
  commutator sparking at, 37
  coils, shunt circuit in, 38
  failure to excite, 38
  armature, repairs to, 39
    broken wire in, 40
  dried by steam, 41
  for charging accumulator, 45
  switching on accumulator, 50
  and accumulator, leads, 50
Dynamos, parallel running of different, 52
  periodicity of, 54
  phase of alternations, 54
    parallel, alternating, 54
  regulation of, 72
  for arcs in parallel and series, 92
  for ship work, 199
  driving aboard ship, 200
Dynamo-room switchboard, 122

EARTH fault at dynamo, 33
  Economical efficiency of the incandescent lamp, 170
Edison-Howell lamp indicator, 68
Efficiency of the incandescent lamp, 170
Electrolyte of accumulator, 44
Electricity supply, meters for, 120
Electromotive force and current for incandescent lamps, 168
Electromotive force, unit of, 184
  required for given lamps, 188
Electrical power, estimation of the, 184
Electric lamps on gas fittings, 196

Electric lamps, for compass, 208
Electric projector for canal work, 209
Erection of dynamos, 13
Error of compass due to lighting of ship, 203
Estimation of the electrical power required, 184
Excitation of dynamo, separate, 2
Excess indicator for accumulator, 49

FAILURE of dynamo to excite, 38
Fall of potential, 127
  loss from, 129
  of accumulator, reserve cells for, 51
Faults, localising dynamo, 32
Faults and contacts, intermittent, 36
Faulty accumulator cell, temperature of, 48
Feeders for parallel wiring, 126
Field magnet, direction of current in, 24
Field coils, overheating of, 29
Fire office, rules of insulation resistance, 183
Fittings and lamps, the, 167
Fittings for the incandescent lamp, 172
  for ships, 208
Fluid insulators, 108
Fluxes for soldering, 195
Focussing arc lamps, 91
Foundations for dynamo, 12
Fuse boards, 166
  Scott's safety, 167
Fuses and cut-out, main, 160
  composition of, 163
  observations respecting, 164
Fusible plates, 162
  plugs and branch fuses, 163
Fusing point of copper and alloys, 163

INDEX.

GAP wire gauge, 149
   Gas fittings, electric lamps on, 196
Gas fittings, insulated points in, 196
   adapters for, 197
Gauge of wire for the circuits, 143
Gauges of wire, table of, 146
Geipel relay, 8
General suggestions respecting jointing, 194
Gimmingham and Fleming's voltmeter, 65
Governors, automatic, attention to, 31
Ground leakage in arc circuit, 104

HAND and automatic regulation of current, 5
Hand regulation of dynamo, 6
Hartnell's shade carrier, 173
Heat and attrition in dynamos, 28
Heated bearings of dynamos, 28
Hedges' double pole switch, 153
   fusible safety plug, 162
   fuse for switch board, 166
Hints to accumulator attendants, 42
   working, respecting accumulator, 46
   respecting dynamo working, 30
Holden hydrometer, 62
Horse-power, electrical, 186
Hot lamp, resistance of, 147
House main switch board, 121
Hydrometers for accumulator, 61
   bead form of, 61
   for ships' cell, 61
   Holden's, 62

IMPEDANCE coil, 115
   construction of, 116
   resistance of, 116

Incandescent lamp fittings, 172
   shade carrier for, 173
   fitting of to gas brackets, 197
Incandescent lamps, in series with arc, 101
   wiring for, 124
   in general use, 167
   nature of, 168
   blackening of the envelopes of, 169
   current absorbed by, 170
   economical efficiency of, 170
Incandescent lighting of ships, 199
   lamp for compass, 208
Indicator, excess, for accumulator, 49
   lamp, Edison-Howell, 68
Indicators, working, 57
Institution of Elec. Eng., rules of, 212
Instruments and tools for jointing, 191
Insulation of dynamo, 12
   of commutator, 18
   of accumulator, 42
   resistance to take a test of, 81
   tests during wiring, 83, 86
Insulators, pole and wall, 107
   fluid, 108
Insulated leads for arc lighting, 110
   tube for piercing wall, 111
Insulated wire, nature of, 150
Insulation, red, significance of, 175
   black   ,,   ,, 175
   resistance rules, 183
Insulated joints for gas fittings, 196
Intermittent faults and contacts, 36

JOHNSON & PHILLIPS' fluid insulator, 108
Jointing conductors, 190

Joint in gutta-percha-covered wire, 191
  in a taped lead, 192
    cable, 194

K ILOWATT, the, 171
  Kilowatt-hour, 171, 187
Key and plug switches, 159

L ACING of belting, 15
  Lamp, Brush's adjustment of, 99
  indicator, Edison-Howell's, 68
  pendants telescopic, 175
  power and area lighted, 167
  switches, 157
  trimming, arc, 93
Lamps arc, focussing, 91
  in parallel, 91
  distance between carbons in, 91
  rules for adjusting, 94
Lamps, bank of, as a Rheostat, 53
  in parallel, 136
  resistance of, 147
  and fittings, the, 167
  incandescent, in general use, 167
    blackening of bulbs of, 169
    current absorbed by, 170
Law of the circuit, Ohm's, 186
Leakage to earth, rough test for, 37
"Lead" in adjustment of brushes, 11
Leads and contacts for accumulators, 50
  running for arc lamps, 102
  rule of the road for, 175
  and branches, T-joint in, 193
Lightning, damages by, 106
  arrester, 106
Lightly insulated leads, 109
Light absorbed by glass envelopes, 167
Localising dynamo faults, 32
  broken conductor, 35

Location of transformer, 118
Long shunt compound winding, 5
Loop circuits, closed, 141
Lubrication of commutator, 18
Lubricators, needle, 26

M AGNET coil, shunt circuit in, 38
Magnetic Voltmeters, 62
Mains and Feeders, planning system of, 111
Main switches, 151
Main switch, double pole, 153
  multiple way, 154
Main fuses and cut-outs, 160
Main lead, T-joint in, 193
Management of dynamos, 12
Materials for jointing, 190
Mechanical test of dynamo, 25
Meters, ampère for station work, 67
  for electricity supply, 120
Methods of running wires, 175
  jointing conductors, 190
Mica for commutator insulation, 18
Mills and factories, running arc leads in, 111
Miscellaneous information, 212
Multiple arc wiring, 125
  series system, the, 134
  way main switches, 154

N AKED leads in arc lighting, 109
Nature of the insulation of wires, 150
  incandescent lamp, 168
Needle lubricators, 26
Neutral point for brushes, 11
  constant position of, 12
New commutator, fitting of, 20
New buildings, wiring of, 179
Notes for dynamo attendants, 26

O BSOLETE, single arc system the, 89

INDEX.   223

Observations respecting fuses, 164
Ohm, the, 185
Ohm's law of the circuit, 186
Overheating of field coils, 29
Overloading of dynamo, 29
Overheated armature, 29

PARALLEL running or shunt dynamos in parallel, 52
    running of alternating dynamos, 54
    arc lamps in, 91
    and series running of arcs, dynamos for, 92
    transformers in, 119
    wiring, 125
    feeders for, 126
    Lamps in, 136
    or parallel series, 139
Paterson's voltmeters, 64
Paterson & Cooper's fuse-plate, 164
Pendants, telescopic, for lamps, 175
Pendant arms for incandescent lamps, 198
Periodic vibration of brushes, 17
    faults of dynamo, 33
Periodicity of alternating dynamos, 54
Personal precautions, 31
Phase of alternating dynamo, 54
Planning system of mains and feeders, 111
Planning of circuits, 140
Plug and key switches, 159
Plugs, fusible branch, 163
Pocket voltmeter, 65
Pole and wall insulators, 107
Portable Wheatstone's bridge, 82
    lamps, attachments of, 173
Potential, fall of, 127
Power and ratio of belt surface, 15
Precautions, personal, 31
Pressure, fall of, 127

Prevention of sulphating in accumulator, 47
Protection, mechanical, of cable, 103

RATE of discharge of accumulator, 46
Ratio of belting surface to power, 15
Red insulation, significance of, 175
Regulation, hand and automatic, of dynamos, 5
    hand, of dynamo, 6
    automatic, of dynamo, 6
    of Brush's dynamo, 7
    of Thomson - Houston's dynamo, 9
    of dynamos to correspond with indicator, 72
    of arcs in series, 92
Regulating coil, arc, single, 89
Relation of electricity to mechanical units, 186
Relay, Geipel's, 8
Repairs to the armature, 39
Reserve cells of accumulator, 51
Resistance for dynamo, artificial, 53
    variable, 73
    and insulation tests, 74
    insulation, taking tests of, 81
    tests, 85
    of lamps in parallel, 147
    unit of, 185
Returning of commutator, 19
Reversing switch, 160
Rheostat, bank of lamps as a, 53
Rheostats, 73
Ring-contact switches, 152
Rope driving of dynamos, 15
Roughness of commutator, treatment of, 19
Rule of the road for wires, 175
Rules of the Inst. of E. E., 212
Running series or shunt dynamos in parallel, 52
    leads for arc lamps, 102

INDEX.

SAFETY fuses for transformers, 120
   plugs, branch, 163
   fuse, Scott's, 167
Scott's safety fuse, 167
Selection of a system, 137
   suggestions respecting, 140
Separate excitation, 2
Series winding of dynamo, 3
   running of arcs, regulation of, 92
   of arc lamps, 92
   multiple wiring, 130
   system, the, 134
Shade carrier, Hartnell's, 173
Ship's cell hydrometer, 61
Ship testing, importance of, 86
   lighting, 199
   work, dynamos for, 199
   dynamo driving, 200
   lighting, accumulators in, 201
"Ship-return," 206
Ship wiring, 206
   fittings, 208
Shocks, electric nature of, 31
Short shunt compound winding, 5
   circuit in magnet coil, 38
   circuiting in accumulator, 48
Shunt winding of dynamo, 4
Singing of alternating arc, 118
Single-regulating coil, arc, 89
   arc system, obsolete, 89
   wire work aboard ships, 206
Sir William Thomson's voltmeters, 65
Soldering, fluxes for, 195
   and tinning, 195
Sparking at commutator, 37
Speeding and belting dynamos, 14
Spot, a, on a commutator, 17
Stranded conductor, joint in, 194
Starting accumulator, 43
Station work, ampèremeters for, 67
Steam in drying dynamo, 41

Steel-yard voltmeters, Thomson's, 64
Suez Canal projector, 209
Suggestions for selecting a system, 140
   respecting jointing, 194
Sulphating of accumulator, 47
   prevention of, 47
Switching in dynamo at right instant, 50
Switch-board and testing work, 52
Switch-room indicators, 57
Switch-board, house main, 121
   for dynamo room and accumulators, 122
Switching arrangements, 151
Switch main, 151
Switches, ring-contact, 152
   double-break, 152
   double-pole, 153
   for accumulator, 156
   branch and lamp, 157
   combined with fuse, 159
   plug and ring, 159
   capacity of, 160
Switch-boards, Hedges' fuse for, 167
System of mains and feeders, planning, 111
   the three-wire, 132
   the series, 134
   the multiple-series, 134
   of wiring, selection of, 137
   distributing-box, 141
   the tree, 142

TABLE of wire gauges, 146
T-joint in a branch lead, 193
Telescopic lamp pendants, 175
Temperature of conductor, highest permissible, 212
Tests for leakage in dynamo, 33
   for internal fault in dynamo, 35

INDEX.

Tests for leakage to earth, 37
Testing and switch-board work, 52
Test-lamps for alternators in parallel, 56
Testing resistance and insulation, 74
   box, 76
   taking conductor resistance, 78
Test of insulation resistance, 81
Tests, insulation and conductivity, during wiring, 83
   for continuity, 84
   resistance, 85
   insulation, 86
   for conductivity of wires, 149
   during wiring, 182
   for compass disturbance due to lighting, 205
Thomson-Houston's regulation of dynamo, 9
   air-blast for dynamo, 10
   lightning arrester, 107
Thomson's gravity voltmeter, 64
   (Sir William) rule for conductors and current, 189
Three-wire system, the, 132
Time curve in lighting, 27
   and current curve, 56
Transformer, nature of, 112
   or converters, 112
   working of, 114
   Thomson-Houston's, 114
   location of, 118
   damp, remedy for, 119
   in parallel and series, 119
   necessity for opening primary circuit of, 120
 . safety fuses for, 120
   working off, 135
Treatment of commutator, 17
Tree system, the, 142
Trimming arc lamps, 93
Trotter's dioptric shade for lamps, 173

UNIT of electromotive force, 184
   of resistance, 185

VIBRATION of dynamo, evil effects of, 13
Volt, the, 184
Voltage required for incandescent lamps, 168
Voltmeter, Cardew's, 57
   accumulator, 59
   magnetic, 62
   Ayrton & Perry's, 62
   calibrating, 63
   Paterson's, 64
   Thomson's gravity, 64
   pocket, 65
   Gimmingham & Fleming's, 65
   Sir William Thomson's, 65
   Ayrton & Perry's spring, 66

WALL insulator, 110
   Watt, the, 186
Wheatstone's bridge testing box, 76
   portable, 82
Winding, series, of dynamo, 3
   of dynamo, shunt, 4
   compound, 5
Wire broken in armature, 40
   splice joint in armature, 41
   gauges, table of, 146
   gauges and gauging, 147
   gauge, gap, 149
   size of, for the circuits, 143
Wires, conductivity of, tests for, 149
   insulated, nature of, 150
   methods of running, 175
   jointing, 190
   and cable leads, difference between, 103
Wiring and fitting for arc light, 88
   for incandescent lamps, 124
   parallel, 125
   the system of, 125
   multiple arc, 125

Wiring series, multiple method of, 130
   various systems of, 136
   selection of a system of, 137
   cleat, 176
   in cases and moulding, 177
   of new buildings, 179
   precautions against damp and short circuits in, 180

Wiring, tests during, 182
   aboard ship, 206
Woodhouse & Rawson's main fuses, 161
Work, switchboard and testing, 52
Working indicators of current, 57
   of transformers, 114
   off transformers, 135

THE END.

PRINTED BY J. S. VIRTUE AND CO., LIMITED, CITY ROAD, LONDON.

THE
DIAMOND
Quick Break
SWITCHES

10,000 IN STOCK

AND   AND

10,000 IN STOCK

CUT-OUTS.

Largest Makers in the World.

PRICES, ETC., FROM

WOODHOUSE & RAWSON
UNITED LIMITED
88, Queen Victoria Street, London, E.C.;
41, Piccadilly, Bradford;
Cornbrook Telegraph Works, Manchester.

ADVERTISEMENTS.

E. S. HINDLEY,
11, Queen Victoria Street, E.C.
WORKS:—BOURTON, DORSET.

VERTICAL STEAM ENGINES.
Portable or Fixed.

With reliable Governors for Electric Light.

HORIZONTAL STEAM ENGINES.
With or without Vertical Boilers.

SIMPLE IN DESIGN.
Economical in use.

Illustrated Catalogues, Estimates, and Detailed Information, Free on application.

FREDERICK SMITH & Co., HALIFAX.
Galvanized Iron Telegraph Wire. All Specifications.

High Conductivity Copper Wire.

Soft—Hard Drawn—Tinned—Stranded.

Also Strip and Rod.

BY THE AUTHOR OF "ELECTRIC LIGHT FITTING."

Third Edition, crown 8vo, 396 pp. 7s. 6d. cloth.

ELECTRIC LIGHT: Its Production and Use.

Embodying Plain Directions for the Treatment of Dynamo-Electric Machines, Batteries, Accumulators, and Electric Lamps. With numerous Illustrations.

By JOHN W. URQUHART.

Third Edition, carefully revised, with large additions.

"The book is by far the best that we have yet met with on the subject." [First edition.]—*Athenæum.*

"This is the third and enlarged edition of a work, concerning which there can be but one opinion... The book may be described as a complete directory of the more important patents, and as a miniature *vade mecum* of the salient facts connected with Electric-lighting."—*Electrician.*

"The whole ground of Electric-lighting is more or less covered—accumulators, transformers, meters, &c., being referred to, illustrated, and explained in a very clear and concise manner."—*Telegraphic Journal.*

CROSBY LOCKWOOD & SON, 7, Stationers' Hall Court, London, E.C.

A New Pocket-Book for Electrical Engineers.

Just published, Pocket Size, 4¾ *by* 3 *inches, price* 5/-, *strongly bound.*

THE
Electrical Engineer's Pocket-Book
OF MODERN RULES, FORMULÆ, TABLES, AND DATA.

BY

H. R. KEMPE, M.Inst.E.E., A.M.Inst.C.E.
TECHNICAL OFFICER, POSTAL TELEGRAPHS; AUTHOR OF "A HANDBOOK OF ELECTRICAL TESTING," ETC., ETC.

With Numerous Illustrations.

EXTRACT FROM PREFACE.

IN this book the Author's aim has been to produce a work which should, as far as possible, truly accord with its title, viz.—a *Pocket Book* of useful "Modern Rules, Formulæ, Tables, and Data," the matter selected being such as is required for daily use in Practical Electrical Engineering, and its various applications to the Industries of the present day. . . . . In its range, the book covers a very wide field of present practice, all matters which possess merely an historical or literary interest being avoided.

The volume is divided into twenty-one sections, as follows: (1) Weights and Measures; (2) Units; (3) Temperature; (4) Electro-Chemistry, Primary Batteries, and Accumulators; (5) Electro-Metallurgy; (6) Currents; (7) Resistance; (8) Capacity; (9) Galvanometers; (10) Fault Testing; (11) Wire; (12) Insulated Wire; (13) Electric Light Leads; (14) Electric Light Dynamos and Motors; (15) Rules and Regulations; (16) Telegraph Apparatus; (17) Telephone; (18) Miscellaneous; (19) Mathematical Tables; (20) Foreign Moneys; (21) Dictionary of Technical Terms.

CROSBY LOCKWOOD & SON, 7, Stationers' Hall Court, London, E.C.

BOOKS FOR
ELECTRICIANS, ELECTRO-METALLURGISTS, &c.

THE STUDENT'S TEXT-BOOK OF ELECTRICITY. By HENRY M. NOAD, Ph.D., F.R.S., F.C.S. New Edition, carefully Revised. With an Introduction and Additional Chapters, by W. H. PREECE, M.I.C.E., Vice-President of the Society of Telegraph Engineers, &c. With 470 Illustrations. Crown 8vo, 12s. 6d. cloth.

"We can recommend Dr. Noad's book for clear style, great range of subject, a good index, and a plethora of woodcuts. Such collections as the present are indispensable."—*Athenæum.*

THE ELEMENTARY PRINCIPLES OF ELECTRIC LIGHTING. By ALAN A. CAMPBELL SWINTON, Associate I.E.E. Second Edition, Enlarged and Revised. With 16 Illustrations. Crown 8vo, 1s. 6d.

"Anyone who desires a short and thoroughly clear exposition of the elementary principles of electric-lighting cannot do better than read this little work."—*Bradford Observer.*

ELECTROPLATING: A Practical Handbook. By J. W. URQUHART, C.E., Author of "Electric Light, its Production and Use," "Electric Light Fitting," &c., &c. Second Edition, thoroughly Revised and Enlarged. With numerous Illustrations. Crown 8vo, 5s. cloth.

"The information given appears to be based on direct personal knowledge. . . . . Its science is sound and the style is always clear."—*Athenæum.*

ELECTROTYPING: The Reproduction and Multiplication of Printing Surfaces and Works of Art by the Electro-deposition of Metals. By J. W. URQUHART, C.E., Author of "Electric Light, its Production and Use," "Electric Light Fitting," &c., &c. Crown 8vo, 5s. cloth.

"The book is thoroughly practical. The reader is, therefore, conducted through the leading laws of electricity, then through the metals used by electrotypers, the apparatus, and the depositing processes, up to the final preparation of the work."—*Art Journal.*

ELECTRO-DEPOSITION: A Practical Treatise on the Electrolysis of Gold, Silver, Copper, Nickel, and other Metals and Alloys. With Descriptions of Voltaic Batteries, Magnet and Dynamo-Electric Machines, Thermopiles, and of the Materials and Processes used in every Department of the Art, and several Chapters on ELECTRO-METALLURGY. By ALEXANDER WATT, Author of "Electro-Metallurgy," &c. With numerous Illustrations. Third Edition, Revised and Enlarged. Crown 8vo, 9s. cloth.

"Evidently written by a practical man who has spent a long period of time in electroplate workshops. The information given respecting the details of workshop manipulation is remarkably complete."—*Nature.*

ELECTRO-METALLURGY: Practically Treated. By ALEXANDER WATT, F.R.S.S.A. Ninth Edition, Revised, with Additional Matter and Illustrations, including the most recent Processes. 12mo, 3s. 6d. cloth boards.

"From this book both amateur and artisan may learn everything necessary for the successful prosecution of electroplating."—*Iron.*

CROSBY LOCKWOOD & SON, 7, Stationers' Hall Court, London, E.C.

PRACTICAL BOOKS FOR ENGINEERS.

THE PRACTICAL ENGINEER'S HANDBOOK. Comprising a Treatise on Modern Engines and Boilers, Marine, Locomotive and Stationary. And containing a large collection of Rules and Practical Data relating to recent Practice in Designing and Constructing all kinds of Engines, Boilers, and other Engineering work. By WALTER S. HUTTON, Civil and Mechanical Engineer, Author of "The Works' Manager's Hand-book for Engineers," &c. With upwards of 370 Illustrations. Third Edition, Revised with Additions. Medium 8vo, nearly 500 pp., price 18s. Strongly bound.

"We have kept it at hand for several weeks, referring to it as occasion arose, and we have not on a single occasion consulted its pages without finding the information of which we were in quest."—*Athenæum.*

THE WORKS' MANAGER'S HANDBOOK OF MODERN RULES, TABLES, AND DATA. For Engineers, Millwrights, and Boiler Makers; Tool Makers, Machinists, and Metal Workers; Iron and Brass Founders, &c. By W. S. HUTTON, Civil and Mechanical Engineer, Author of "The Practical Engineer's Handbook." Fourth Edition, carefully Revised, with Additions. In one handsome Volume, medium 8vo, price 15s. Strongly bound.

"The author treats every subject from the point of view of one who has collected workshop notes for application in workshop practice, rather than from the theoretical or literary aspect. The volume contains a great deal of that kind of information which is gained only by practical experience, and is seldom written in books."—*Engineer.*

THE PRACTICAL MECHANIC'S WORKSHOP COMPANION. Comprising a great variety of the most useful Rules and Formulæ in Mechanical Science, with numerous Tables of Practical Data and Calculated Results for Facilitating Mechanical Operations. By WILLIAM TEMPLETON. Sixteenth Edition, Revised, Modernised, and considerably Enlarged by WALTER S. HUTTON, C.E., Author of "The Works' Manager's Handbook." Fcap. 8vo, nearly 500 pp., with 8 Plates and upwards of 250 Illustrative Diagrams, 6s. Strongly bound for workshop or pocket wear and tear.

"In its modernised form Hutton's 'Templeton' should have a wide sale, for it contains much valuable information which the mechanic will often find of use, and not a few tables and notes which he might look for in vain in other works."—*English Mechanic.*

TABLES, MEMORANDA, AND CALCULATED RESULTS, FOR MECHANICS, ENGINEERS, ARCHITECTS, BUILDERS, &c. Selected and arranged by FRANCIS SMITH. Fourth Edition, Revised and Enlarged, 250 pp., waistcoat-pocket size, 1s. 6d. limp leather.

"It would, perhaps, be as difficult to make a small pocket-book selection of notes and formulæ to suit ALL engineers as it would be to make a universal medicine; but Mr. Smith's waistcoat-pocket collection may be looked upon as a successful attempt."
—*Engineer.*
"A veritable pocket treasury of knowledge."—*Iron.*

CROSBY LOCKWOOD & SON, 7, Stationers' Hall Court, London, E.C.

Crown 8vo, over 600 pages, price **4/-** *cloth.*

LOCKWOOD'S
Builders' and Contractors' Price Book
FOR 1891.

ENTIRELY RE-WRITTEN, RE-MODELLED, AND MUCH ENLARGED.

THIS old-established STANDARD PRICE BOOK FOR THE BUILDING TRADE, having been entirely Re-written, Re-modelled, and much Enlarged under the supervision of the Editor, Mr. FRANCIS MILLER, A.R.I.B.A., will be found to contain, in the most convenient form for reference, full details of the most **Recent Improved Methods of Construction,** and every special feature of Modern Building—such as

ELECTRIC LIGHTING,
VENTILATION, and the NEWEST SANITARY APPLIANCES.

THE PHŒNIX FIRE OFFICE RULES

FOR

ELECTRIC LIGHT INSTALLATIONS
AND
ELECTRICAL POWER INSTALLATIONS

Are also given in full.

In order to enhance the usefulness and convenience of the PRICE-BOOK for the purposes of daily reference, the entire work has been re-set in a new and elegant type. With these improvements, the work will undoubtedly maintain its position as **the Handiest and Completest Price-Book for Builders and Contractors.**

CROSBY LOCKWOOD & SON, 7, Stationers' Hall Court, London, E.C.

7, Stationers' Hall Court, London, E.C.
March, 1891.

A

CATALOGUE OF BOOKS

INCLUDING NEW AND STANDARD WORKS IN
ENGINEERING: CIVIL, MECHANICAL, AND MARINE,
MINING AND METALLURGY,
ELECTRICITY AND ELECTRICAL ENGINEERING,
ARCHITECTURE AND BUILDING,
INDUSTRIAL AND DECORATIVE ARTS, SCIENCE, TRADE,
AGRICULTURE, GARDENING,
LAND AND ESTATE MANAGEMENT, LAW, &c.

PUBLISHED BY

CROSBY LOCKWOOD & SON.

MECHANICAL ENGINEERING, etc.

New Manual for Practical Engineers.
THE PRACTICAL ENGINEER'S HAND-BOOK. Comprising a Treatise on Modern Engines and Boilers: Marine, Locomotive and Stationary. And containing a large collection of Rules and Practical Data relating to recent Practice in Designing and Constructing all kinds of Engines, Boilers, and other Engineering work. The whole constituting a comprehensive Key to the Board of Trade and other Examinations for Certificates of Competency in Modern Mechanical Engineering. By WALTER S. HUTTON, Civil and Mechanical Engineer, Author of "The Works' Manager's Handbook for Engineers," &c. With upwards of 370 Illustrations. Third Edition, Revised, with Additions. Medium 8vo, nearly 500 pp., price 18s. Strongly bound.

☞ *This work is designed as a companion to the Author's* "WORKS' MANAGER'S HAND-BOOK." *It possesses many new and original features, and contains, like its predecessor, a quantity of matter not originally intended for publication, but collected by the author for his own use in the construction of a great variety of modern engineering work.*

The information is given in a condensed and concise form, and is illustrated by upwards of 370 Woodcuts; and comprises a quantity of tabulated matter of great value to all engaged in designing, constructing, or estimating for ENGINES, BOILERS *and* OTHER ENGINEERING WORK.

*** OPINIONS OF THE PRESS.

" We have kept it at hand for several weeks, referring to it as occasion arose, and we have not on a single occasion consulted its pages without fi ling the information of which we were in quest." —*Athenæum.*

" A thoroughly good practical handbook, whi h no engineer can go through without learning something that will be of service to him."—*Marine Engineer.*

" An excellent book of reference for engineers, and a valuable text-book for students of engineering."—*Scotsman.*

" This valuable manual embodies the results and experience of the leading authorities on mechanical engineering."—*Building News.*

" The author has collected together a surprising quantity of rules and practical data, and has shown much judgment in the selections he has made. . . . There is no doubt that this book is one of the most useful of its kind published, and will be a very popular compendium."—*Engineer.*

" A mass of information, set down in simple language, and in such a form that it can be easily referred to at any time. The matter is uniformly good and well chosen, and is greatly elucidated by the illustrations. The book will find its way on to most engineers' shelves, where it will rank as one of the most useful books of reference."—*Practical Engineer.*

" Full of useful information, and should be found on the office shelf of all practical engineers.' —*English Mechanic.*

B

Handbook for Works' Managers.

THE WORKS' MANAGER'S HANDBOOK OF MODERN RULES, TABLES, AND DATA. For Engineers, Millwrights, and Boiler Makers; Tool Makers, Machinists, and Metal Workers; Iron and Brass Founders, &c. By W. S. HUTTON, Civil and Mechanical Engineer, Author of "The Practical Engineer's Handbook." Fourth Edition, carefully Revised, and partly Re-written. In One handsome Volume, medium 8vo, price 15s. strongly bound. [*Just published.*

☞ *The Author having compiled Rules and Data for his own use in a great variety of modern engineering work, and having found his notes extremely useful, decided to publish them—revised to date—believing that a practical work, suited to the* DAILY REQUIREMENTS OF MODERN ENGINEERS, *would be favourably received.*

In the Third Edition, the following among other additions have been made, viz.: Rules for the Proportions of Riveted Joints in Soft Steel Plates, the Results of Experiments by PROFESSOR KENNEDY *for the Institution of Mechanical Engineers—Rules for the Proportions of Turbines—Rules for the Strength of Hollow Shafts of Whitworth's Compressed Steel, &c.*

*** OPINIONS OF THE PRESS.

"The author treats every subject from the point of view of one who has collected workshop notes for application in workshop practice, rather than from the theoretical or literary aspect. The volume contains a great deal of that kind of information which is gained only by practical experience, and is seldom written in books."—*Engineer.*

"The volume is an exceedingly useful one, brimful with engineers' notes, memoranda, and rules, and well worthy of being on every mechanical engineer's bookshelf."—*Mechanical World.*

"The information is precisely that likely to be required in practice. . . . The work forms a desirable addition to the library not only of the works manager, but of anyone connected with general engineering."—*Mining Journal.*

"A formidable mass of facts and figures, readily accessible through an elaborate Index . . . . Such a volume will be found absolutely necessary as a book of reference in all sorts of 'works' connected with the metal trades."—*Ryland's Iron Trades Circular.*

"Brimful of useful information, stated in a concise form, Mr. Hutton's books have met a pressing want among engineers. The book must prove extremely useful to every practical man possessing a copy."—*Practical Engineer.*

Practical Treatise on Modern Steam-Boilers.

STEAM-BOILER CONSTRUCTION. A Practical Handbook for Engineers, Boiler-Makers, and Steam Users. Containing a large Collection of Rules and Data relating to the Design, Construction, and Working of Modern Stationary, Locomotive, and Marine Steam-Boilers. By WALTER S. HUTTON, Civil and Mechanical Engineer, Author of "The Works' Manager's Handbook," "The Practical Engineer's Handbook," &c. With upwards of 300 Illustrations.

☞ *This work is written in the same style as Mr. Hutton's other practical Handbooks, which it is intended to supplement. It is in active preparation and will, it is expected, be ready in April.*

"The Modernised Templeton."

THE PRACTICAL MECHANIC'S WORKSHOP COMPANION. Comprising a great variety of the most useful Rules and Formulæ in Mechanical Science, with numerous Tables of Practical Data and Calculated Results for Facilitating Mechanical Operations. By WILLIAM TEMPLETON, Author of "The Engineer's Practical Assistant," &c. &c. Sixteenth Edition, Revised, Modernised, and considerably Enlarged by WALTER S. HUTTON, C.E., Author of "The Works' Manager's Handbook," "The Practical Engineer's Handbook," &c. Fcap. 8vo, nearly 500 pp., with Eight Plates and upwards of 250 Illustrative Diagrams, 6s., strongly bound for workshop or pocket wear and tear. [*Just published.*

*** OPINIONS OF THE PRESS.

"In its modernised form Hutton's 'Templeton' should have a wide sale, for it contains much valuable information which the mechanic will often find of use, and not a few tables and notes which he might look for in vain in other works. This modernised edition will be appreciated by all who have learned to value the original editions of 'Templeton.'"—*English Mechanic.*

"It has met with great success in the engineering workshop, as we can testify; and there are a great many men who, in a great measure, owe their rise in life to this little book."—*Building News.*

"This familiar text-book—well known to all mechanics and engineers—is of essential service to the every-day requirements of engineers, millwrights, and the various trades connected with engineering and building. The new modernised edition is worth its weight in gold."—*Building News.* (Second Notice.)

"This well-known and largely used book contains information, brought up to date, of the sort so useful to the foreman and draughtsman. So much fresh information has been introduced as to constitute it practically a new book. It will be largely used in the office and workshop."—*Mechanical World.*

"The publishers wisely entrusted the task of revision of this popular, valuable and useful book to Mr. Hutton, than whom a more competent man they could not have found."—*Iron.*

MECHANICAL ENGINEERING, etc. 3

Stone-working Machinery.
STONE-WORKING MACHINERY, and the Rapid and Economical Conversion of Stone. With Hints on the Arrangement and Management of Stone Works. By M. POWIS BALE, M.I.M.E. With Illusts. Crown 8vo, 9s.
"The book should be in the hands of every mason or student of stone-work."—*Colliery Guardian.*
"It is in every sense of the word a standard work upon a subject which the author is fully competent to deal exhaustively with."—*Builder's Weekly Reporter.*
"A capital handbook for all who manipulate stone for building or ornamental purposes."—*Machinery Market.*

Pump Construction and Management.
PUMPS AND PUMPING: A Handbook for Pump Users. Being Notes on Selection, Construction and Management. By M. POWIS BALE, M.I.M.E., Author of "Woodworking Machinery," "Saw Mills," &c. Crown 8vo, 2s. 6d. cloth. [*Just published.*
"The matter is set forth as concisely as possible. In fact, condensation rather than diffuseness has been the author's aim throughout; yet he does not seem to have omitted anything likely to be of use."—*Journal of Gas Lighting.*
"Thoroughly practical and simply and clearly written."—*Glasgow Herald.*

Turning.
LATHE-WORK: A Practical Treatise on the Tools, Appliances, and Processes employed in the Art of Turning. By PAUL N. HASLUCK. Fourth Edition, Revised and Enlarged. Cr. 8vo, 5s. cloth. [*Just published.*
"Written by a man who knows, not only how work ought to be done, but who also knows how to do it, and how to convey his knowledge to others. To all turners this book would be valuable."—*Engineering.*
"We can safely recommend the work to young engineers. To the amateur it will simply be invaluable. To the student it will convey a great deal of useful information."—*Engineer.*
"A compact, succinct, and handy guide to lathe-work did not exist in our language until Mr. Hasluck, by the publication of this treatise, gave the turner a true *vade-mecum.*"—*House Decorator*

Screw-Cutting.
SCREW THREADS: And Methods of Producing Them. With Numerous Tables, and complete directions for using Screw-Cutting Lathes. By PAUL N. HASLUCK, Author of "Lathe-Work," &c. With Fifty Illustrations. Third Edition, Revised and Enlarged. Waistcoat-pocket size, 1s. 6d. cloth. [*Just published.*
"Full of useful information, hints and practical criticism. Taps, dies and screwing-tools generally are illustrated and their action described."—*Mechanical World.*
"It is a complete compendium of all the details of the screw cutting lathe; in fact a *multum-in-parvo* on all the subjects it treats upon."—*Carpenter and Builder.*

Smith's Tables for Mechanics, etc.
TABLES, MEMORANDA, AND CALCULATED RESULTS, FOR MECHANICS, ENGINEERS, ARCHITECTS, BUILDERS, etc. Selected and Arranged by FRANCIS SMITH. Fifth Edition, thoroughly Revised and Enlarged, with a New Section of ELECTRICAL TABLES, FORMULÆ, and MEMORANDA. Waistcoat-pocket size, 1s. 6d. limp leather. [*Just published.*
"It would, perhaps, be as difficult to make a small pocket-book selection of notes and formulæ to suit ALL engineers as it would be to make a universal medicine; but Mr. Smith's waistcoat-pocket collection may be looked upon as a successful attempt."—*Engineer.*
"The best example we have ever seen of 250 pages of useful matter packed into the dimensions of a card-case."—*Building News.* "A veritable pocket treasury of knowledge."—*Iron.*

Engineer's and Machinist's Assistant.
THE ENGINEER'S, MILLWRIGHT'S, and MACHINIST'S PRACTICAL ASSISTANT. A collection of Useful Tables, Rules and Data. By WILLIAM TEMPLETON. 7th Edition, with Additions. 18mo, 2s. 6d. cloth.
"Occupies a foremost place among books of this kind. A more suitable present to an apprentice to any of the mechanical trades could not possibly be made."—*Building News.*
"A deservedly popular work, it should be in the 'drawer' of every mechanic."—*English Mechanic.*

Iron and Steel.
"IRON AND STEEL": A Work for the Forge, Foundry, Factory, and Office. Containing ready, useful, and trustworthy Information for Iron-masters and their Stock-takers; Managers of Bar, Rail, Plate, and Sheet Rolling Mills; Iron and Metal Founders; Iron Ship and Bridge Builders; Mechanical, Mining, and Consulting Engineers; Architects, Contractors, Builders, and Professional Draughtsmen. By CHARLES HOARE, Author of "The Slide Rule," &c. Eighth Edition, Revised throughout and considerably Enlarged. 32mo, 6s. leather.
"For comprehensiveness the book has not its equal."—*Iron.*
"One of the best of the pocket books."—*English Mechanic.*
"We cordially recommend this book to those engaged in considering the details of all kinds of iron and steel works."—*Naval Science.*

Engineering Construction.

PATTERN-MAKING: A Practical Treatise, embracing the Main Types of Engineering Construction, and including Gearing, both Hand and Machine made, Engine Work, Sheaves and Pulleys, Pipes and Columns, Screws, Machine Parts, Pumps and Cocks, the Moulding of Patterns in Loam and Greensand, &c., together with the methods of Estimating the weight of Castings; to which is added an Appendix of Tables for Workshop Reference. By a FOREMAN PATTERN MAKER. With upwards of Three Hundred and Seventy Illustrations. Crown 8vo, 7s. 6d. cloth.

"A well-written technical guide, evidently written by a man who understands and has practised what he has written about. . . . We cordially recommend it to engineering students, young journeymen, and others desirous of being initiated into the mysteries of pattern-making."—*Builder.*
"We can confidently recommend this comprehensive treatise."—*Building News.*
"Likely to prove a welcome guide to many workmen, especially to draughtsmen who have lacked a training in the shops, pupils pursuing their practical studies in our factories, and to employers and managers in engineering works."—*Hardware Trade Journal.*
"More than 370 illustrations help to explain the text, which is, however, always clear and explicit, thus rendering the work an excellent *vade mecum* for the apprentice who desires to become master of his trade."—*English Mechanic.*

Dictionary of Mechanical Engineering Terms.

LOCKWOOD'S DICTIONARY OF TERMS USED IN THE PRACTICE OF MECHANICAL ENGINEERING, embracing those current in the Drawing Office, Pattern Shop, Foundry, Fitting, Turning, Smith's and Boiler Shops, &c. &c. Comprising upwards of 6,000 Definitions. Edited by A FOREMAN PATTERN-MAKER, Author of "Pattern Making." Crown 8vo, 7s. 6d. cloth.

"Just the sort of handy dictionary required by the various trades engaged in mechanical engineering. The practical engineering pupil will find the book of great value in his studies, and every foreman engineer and mechanic should have a copy."—*Building News.*
"After a careful examination of the book, and trying all manner of words, we think that the engineer will here find all he is likely to require. It will be largely used."—*Practical Engineer.*
"One of the most useful books which can be presented to a mechanic or student."—*English Mechanic.*
"Not merely a dictionary, but, to a certain extent, also a most valuable guide. It strikes us as a happy idea to combine with a definition of the phrase useful information on the subject of which it treats."—*Machinery Market.*
"No word having connection with any branch of constructive engineering seems to be omitted. No more comprehensive work has been, so far, issued."—*Knowledge.*
"We strongly commend this useful and reliable adviser to our friends in the workshop, and to students everywhere."—*Colliery Guardian.*

Steam Boilers.

A TREATISE ON STEAM BOILERS: Their Strength, Construction, and Economical Working. By ROBERT WILSON, C.E. Fifth Edition. 12mo, 6s. cloth.

"The best treatise that has ever been published on steam boilers."—*Engineer.*
"The author shows himself perfect master of his subject, and we heartily recommend all employing steam power to possess themselves of the work."—*Ryland's Iron Trade Circular.*

Boiler Chimneys.

BOILER AND FACTORY CHIMNEYS; Their Draught-Power and Stability. With a Chapter on *Lightning Conductors*. By ROBERT WILSON, A.I.C.E., Author of "A Treatise on Steam Boilers," &c. Second Edition. Crown 8vo, 3s. 6d. cloth.

"Full of useful information, definite in statement, and thoroughly practical in treatment."—*The Local Government Chronicle.*
"A valuable contribution to the literature of scientific building."—*The Builder.*

Boiler Making.

THE BOILER-MAKER'S READY RECKONER & ASSISTANT. With Examples of Practical Geometry and Templating, for the Use of Platers, Smiths and Riveters. By JOHN COURTNEY. Edited by D. K. CLARK, M.I.C.E. Third Edition, 480 pp., with 140 Illusts. Fcap. 8vo, 7s. half-bound.

"A most useful work. . . . No workman or apprentice should be without this book."—*Iron Trade Circular.*
"Boiler-makers will readily recognise the value of this volume. . . . The tables are clearly printed, and so arranged that they can be referred to with the greatest facility, so that it cannot be doubted that they will be generally appreciated and much used."—*Mining Journal.*

Warming.

HEATING BY HOT WATER; with Information and Suggestions on the best Methods of Heating Public, Private and Horticultural Buildings. By WALTER JONES. With upwards of 50 Illustrations, crown 8vo, 2s. cloth.

"We confidently recommend all interested in heating by hot water to secure a copy of this valuable little treatise."—*The Plumber and Decorator.*

Steam Engine.

TEXT-BOOK ON THE STEAM ENGINE. With a Supplement on Gas Engines, and PART II. ON HEAT ENGINES. By T. M. GOODEVE, M.A., Barrister-at-Law, Professor of Mechanics at the Normal School of Science and the Royal School of Mines; Author of "The Principles of Mechanics," "The Elements ot Mechanism," &c. Eleventh Edition, Enlarged. With numerous Illustrations. Crown 8vo, 6s. cloth. [*Just published.*]

"Professor Goodeve has given us a treatise on the steam engine which will bear comparison with anything written by Huxley or Maxwell, and we can award it no higher praise."—*Engineer.*
"Mr. Goodeve's text-book is a work of which every young engineer should possess himself."—*Mining Journal.*
"Essentially practical in its aim. The manner of exposition leaves nothing to be desired."—*Scotsman.*

Gas Engines.

ON GAS-ENGINES. Being a Reprint, with some Additions, of the Supplement to the *Text-book on the Steam Engine*, by T. M. GOODEVE, M.A. Crown 8vo, 2s. 6d. cloth.

"Like all Mr. Goodeve's writings, the present is no exception in point of general excellence. It is a valuable little volume."—*Mechanical World.*

Steam.

THE SAFE USE OF STEAM. Containing Rules for Unprofessional Steam-users. By an ENGINEER. Sixth Edition. Sewed, 6d.

"If steam-users would but learn this little book by heart boiler explosions would become sensations by their rarity."—*English Mechanic.*

Office Book for Mechanical Engineers.

THE MECHANICAL ENGINEER'S REFERENCE BOOK, for Machine and Boiler Construction. In Two Parts. Part I. GENERAL ENGINEERING DATA. Part II. BOILER CONSTRUCTION. With 48 Plates and numerous Illustrations. By NELSON FOLEY, M.I.N.A. Folio, half-bound. Price £5 5s. [*Nearly ready.*]

Coal and Speed Tables.

A POCKET BOOK OF COAL AND SPEED TABLES, for *Engineers and Steam-users.* By NELSON FOLEY, Author of "Boiler Construction." Pocket-size, 3s. 6d. cloth; 4s. leather.

"These tables are designed to meet the requirements of every-day use; they are of sufficient scope for most practical purposes, and may be commended to engineers and users of steam."—*Iron.*
"This pocket-book well merits the attention of the practical engineer. Mr. Foley has compiled a very useful set of tables, the information contained in which is frequently required by engineers, coal consumers and users of steam."—*Iron and Coal Trades Review.*

Fire Engineering.

FIRES, FIRE-ENGINES, AND FIRE-BRIGADES. With a History of Fire-Engines, their Construction, Use, and Management; Remarks on Fire-Proof Buildings, and the Preservation of Life from Fire; Statistics of the Fire Appliances in English Towns; Foreign Fire Systems; Hints on Fire Brigades, &c. &c. By CHARLES F. T. YOUNG, C.E. With numerous Illustrations, 544 pp., demy 8vo, £1 4s. cloth.

"To such of our readers as are interested in the subject of fires and fire apparatus, we can most heartily commend this book. It is really the only English work we now have upon the subject."—*Engineering.*
"It displays much evidence of careful research; and Mr. Young has put his facts neatly together. It is evident enough that his acquaintance with the practical details ot the construction of steam fire engines, old and new, and the conditions with which it is necessary they should comply, is accurate and full."—*Engineer.*

Estimating for Engineering Work, &c.

ENGINEERING ESTIMATES, COSTS AND ACCOUNTS: A Guide to Commercial Engineering. With numerous Examples of Estimates and Costs of Millwright Work, Miscellaneous Productions, Steam Engines and Steam Boilers; and a Section on the Preparation of Costs Accounts. By A GENERAL MANAGER. Demy 8vo, 12s. cloth. [*Just published.*]

"This is an excellent and very useful book, covering subject-matter in constant requisition in every factory and workshop. . . . The book is invaluable, not only to the young engineer but also to the estimate department of every works."—*Builder.*
"This book bears on every page evidence that it has been prepared by an engineer accustomed to the work, and is no mere compilation, but contains a mass of valuable information o, a kind useful even to experienced engineers."—*Practical Engineer.*
"We accord the work unqualified praise. The information is given in a plain, straightforward manner, and bears throughout evidence of the intimate practical acquaintance of the author with every phrase of commercial engineering."—*Mechanical World.*

THE POPULAR WORKS OF MICHAEL REYNOLDS
("THE ENGINE DRIVER'S FRIEND").

Locomotive-Engine Driving.
LOCOMOTIVE-ENGINE DRIVING: A Practical Manual for Engineers in charge of Locomotive Engines. By MICHAEL REYNOLDS, Member of the Society of Engineers, formerly Locomotive Inspector L. B. and S. C. R. Eighth Edition. Including a KEY TO THE LOCOMOTIVE ENGINE. With Illustrations and Portrait of Author. Crown 8vo, 4s. 6d. cloth.

"Mr. Reynolds has supplied a want, and has supplied it well. We can confidently recommend the book, not only to the practical driver, but to everyone who takes an interest in the performance of locomotive engines."—*The Engineer.*

"Mr. Reynolds has opened a new chapter in the literature of the day. This admirable practical treatise, of the practical utility of which we have to speak in terms of warm commendation."—*Athenæum.*

"Evidently the work of one who knows his subject thoroughly."—*Railway Service Gazette.*

"Were the cautions and rules given in the book to become part of the every-day working of our engine-drivers, we might have fewer distressing accidents to deplore."—*Scotsman.*

Stationary Engine Driving.
STATIONARY ENGINE DRIVING: A Practical Manual for Engineers in charge of Stationary Engines. By MICHAEL REYNOLDS. Fourth Edition, Enlarged. With Plates and Woodcuts. Crown 8vo, 4s. 6d. cloth.

"The author is thoroughly acquainted with his subjects, and his advice on the various points treated is clear and practical. . . . He has produced a manual which is an exceedingly useful one for the class for whom it is specially intended."—*Engineering.*

"Our author leaves no stone unturned. He is determined that his readers shall not only know something about the stationary engine, but all about it."—*Engineer.*

"An engineman who has mastered the contents of Mr. Reynolds's book will require but little actual experience with boilers and engines before he can be trusted to look after them."—*English Mechanic.*

The Engineer, Fireman, and Engine-Boy.
THE MODEL LOCOMOTIVE ENGINEER, FIREMAN, and ENGINE-BOY. Comprising a Historical Notice of the Pioneer Locomotive Engines and their Inventors. By MICHAEL REYNOLDS. With numerous Illustrations and a fine Portrait of George Stephenson. Crown 8vo, 4s. 6d. cloth.

"From the technical knowledge of the author it will appeal to the railway man of to-day more forcibly than anything written by Dr. Smiles. . . . The volume contains information of a technical kind, and facts that every driver should be familiar with."—*English Mechanic.*

"We should be glad to see this book in the possession of everyone in the kingdom who has ever laid, or is to lay, hands on a locomotive engine."—*Iron.*

Continuous Railway Brakes.
CONTINUOUS RAILWAY BRAKES: A Practical Treatise on the several Systems in Use in the United Kingdom; their Construction and Performance. With copious Illustrations and numerous Tables. By MICHAEL REYNOLDS. Large crown 8vo, 9s. cloth.

"A popular explanation of the different brakes. It will be of great assistance in forming public opinion, and will be studied with benefit by those who take an interest in the brake."—*English Mechanic.*

"Written with sufficient technical detail to enable the principle and relative connection of the various parts of each particular brake to be readily grasped."—*Mechanical World.*

Engine-Driving Life.
ENGINE-DRIVING LIFE: Stirring Adventures and Incidents in the Lives of Locomotive-Engine Drivers. By MICHAEL REYNOLDS. Second Edition, with Additional Chapters. Crown 8vo, 2s. cloth.

"From first to last perfectly fascinating. Wilkie Collins's most thrilling conceptions are thrown into the shade by true incidents, endless in their variety, related in every page."—*North British Mail.*

"Anyone who wishes to get a real insight into railway life cannot do better than read 'Engine-Driving Life' for himself; and if he once take it up he will find that the author's enthusiasm and real love of the engine-driving profession will carry him on till he has read every page."—*Saturday Review.*

Pocket Companion for Enginemen.
THE ENGINEMAN'S POCKET COMPANION AND PRACTICAL EDUCATOR FOR ENGINEMEN, BOILER ATTENDANTS, AND MECHANICS. By MICHAEL REYNOLDS. With Forty-five Illustrations and numerous Diagrams. Second Edition, Revised. Royal 18mo, 3s. 6d., strongly bound for pocket wear.

"This admirable work is well suited to accomplish its object, being the honest workmanship of a competent engineer."—*Glasgow Herald.*

"A most meritorious work, giving in a succinct and practical form all the information an engine-minder desirous of mastering the scientific principles of his daily calling would require."—*Miller.*

"A boon to those who are striving to become efficient mechanics."—*Daily Chronicle.*

CIVIL ENGINEERING, SURVEYING, etc.

French-English Glossary for Engineers, etc.
A POCKET GLOSSARY of TECHNICAL TERMS: ENGLISH-FRENCH, FRENCH-ENGLISH; with Tables suitable for the Architectural, Engineering, Manufacturing and Nautical Professions. By JOHN JAMES FLETCHER, Engineer and Surveyor. 200 pp. Waistcoat-pocket size, 1s. 6d., limp leather.

"It ought certainly to be in the waistcoat-pocket of every professional man."—*Iron.*
"It is a very great advantage for readers and correspondents in France and England to have so large a number of the words relating to engineering and manufacturers collected in a liliputian volume. The little book will be useful both to students and travellers."—*Architect.*
"The glossary of terms is very complete, and many of the tables are new and well arranged. We cordially commend the book."—*Mechanical World.*

Portable Engines.
THE PORTABLE ENGINE; ITS CONSTRUCTION AND MANAGEMENT. A Practical Manual for Owners and Users of Steam Engines generally. By WILLIAM DYSON WANSBROUGH. With 90 Illustrations. Crown 8vo, 3s. 6d. cloth.

"This is a work of value to those who use steam machinery. . . . Should be read by every one who has a steam engine, on a farm or elsewhere."—*Mark Lane Express.*
"We cordially commend this work to buyers and owners of steam engines, and to those who have to do with their construction or use."—*Timber Trades Journal.*
"Such a general knowledge of the steam engine as Mr. Wansbrough furnishes to the reader should be acquired by all intelligent owners and others who use the steam engine."—*Building News.*
"An excellent text-book of this useful form of engine, which describes with all necessary minuteness the details of the various devices. . . ' The Hints to Purchasers' contain a good deal of commonsense and practical wisdom."—*English Mechanic.*

CIVIL ENGINEERING, SURVEYING, etc.

MR. HUMBER'S IMPORTANT ENGINEERING BOOKS.

The Water Supply of Cities and Towns.
A COMPREHENSIVE TREATISE on the WATER-SUPPLY OF CITIES AND TOWNS. By WILLIAM HUMBER, A-M.Inst.C.E., and M. Inst. M.E., Author of "Cast and Wrought Iron Bridge Construction," &c. &c. Illustrated with 50 Double Plates, 1 Single Plate, Coloured Frontispiece, and upwards of 250 Woodcuts, and containing 400 pages of Text. Imp. 4to, £6 6s. elegantly and substantially half-bound in morocco.

List of Contents.

I. Historical Sketch of some of the means that have been adopted for the Supply of Water to Cities and Towns.—II. Water and the Foreign Matter usually associated with it.—III. Rainfall and Evaporation.—IV. Springs and the water-bearing formations of various districts.—V. Measurement and Estimation of the flow of Water—VI. On the Selection of the Source of Supply.—VII. Wells.—VIII. Reservoirs.—IX. The Purification of Water.—X. Pumps.— XI. Pumping Machinery. — XII. Conduits.—XIII. Distribution of Water.—XIV. Meters, Service Pipes, and House Fittings.—XV. The Law and Economy of Water Works. XVI. Constant and Intermittent Supply.—XVII. Description of Plates. — Appendices, giving Tables of Rates of Supply, Velocities, &c. &c., together with Specifications of several Works illustrated, among which will be found: Aberdeen, Bideford, Canterbury, Dundee, Halifax, Lambeth, Rotherham, Dublin, and others.

"The most systematic and valuable work upon water supply hitherto produced in English, or in any other language. . . . Mr. Humber's work is characterised almost throughout by an exhaustiveness much more distinctive of French and German than of English technical treatises."—*Engineer.*
"We can congratulate Mr. Humber on having been able to give so large an amount of information on a subject so important as the water supply of cities and towns. The plates, fifty in number, are mostly drawings of executed works, and alone would have commanded the attention of every engineer whose practice may lie in this branch of the profession."—*Builder.*

Cast and Wrought Iron Bridge Construction.
A COMPLETE AND PRACTICAL TREATISE ON CAST AND WROUGHT IRON BRIDGE CONSTRUCTION, including Iron Foundations. In Three Parts—Theoretical, Practical, and Descriptive. By WILLIAM HUMBER, A.M.Inst.C.E., and M.Inst.M.E. Third Edition, Revised and much improved, with 115 Double Plates (20 of which now first appear in this edition), and numerous Additions to the Text. In Two Vols., imp. 4to, £6 16s. 6d. half-bound in morocco.

"A very valuable contribution to the standard literature of civil engineering. In addition to elevations, plans and sections, large scale details are given which very much enhance the instructive worth of those illustrations."—*Civil Engineer and Architect's Journal.*
"Mr. Humber's stately volumes, lately issued—in which the most important bridges erected during the last five years, under the direction of the late Mr. Brunel, Sir W. Cubitt, Mr. Hawkshaw, Mr. Page, Mr. Fowler, Mr. Hemans, and others among our most eminent engineers, are drawn and specified in great detail."—*Engineer*

8   *CROSBY LOCKWOOD & SON'S CATALOGUE.*

MR. HUMBER'S GREAT WORK ON MODERN ENGINEERING.

Complete in Four Volumes, imperial 4to, price £12 12s., half-morocco. Each Volume sold separately as follows:—

A RECORD OF THE PROGRESS OF MODERN ENGINEERING. FIRST SERIES. Comprising Civil, Mechanical, Marine, Hydraulic, Railway, Bridge, and other Engineering Works, &c. By WILLIAM HUMBER, A-M.Inst.C.E., &c. Imp. 4to, with 36 Double Plates, drawn to a large scale, Photographic Portrait of John Hawkshaw, C.E., F.R.S., &c., and copious descriptive Letterpress, Specifications, &c., £3 3s. half-morocco.

List of the Plates and Diagrams.

Victoria Station and Roof, L. B. & S. C. R. (8 plates); Southport Pier (2 plates); Victoria Station and Roof, L. C. & D. and G. W. R. (6 plates); Roof of Cremorne Music Hall; Bridge over G. N. Railway; Roof of Station, Dutch Rhenish Rail (2 plates); Bridge over the Thames, West London Extension Railway (5 plates); Armour Plates: Suspension Bridge, Thames (4 plates); The Allen Engine; Suspension Bridge, Avon (3 plates); Underground Railway (3 plates).

"Handsomely lithographed and printed. It will find favour with many who desire to preserve in a permanent form copies of the plans and specifications prepared for the guidance of the contractors for many important engineering works."—*Engineer.*

HUMBER'S RECORD OF MODERN ENGINEERING. SECOND SERIES. Imp. 4to, with 36 Double Plates, Photographic Portrait of Robert Stephenson, C.E., M.P., F.R.S., &c., and copious descriptive Letterpress, Specifications, &c., £3 3s. half-morocco.

List of the Plates and Diagrams.

Birkenhead Docks, Low Water Basin (15 plates); Charing Cross Station Roof, C. C. Railway (3 plates); Digswell Viaduct, Great Northern Railway; Robbery Wood Viaduct, Great Northern Railway; Iron Permanent Way; Clydach Viaduct, Merthyr, Tredegar, and Abergavenny Railway; Ebbw Viaduct, Merthyr, Tredegar, and Abergavenny Railway; College Wood Viaduct, Cornwall Railway; Dublin Winter Palace Roof (3 plates); Bridge over the Thames, L. C. & D. Railway (6 plates); Albert Harbour, Greenock (4 plates).

"Mr. Humber has done the profession good and true service, by the fine collection of examples he has here brought before the profession and the public."—*Practical Mechanic's Journal.*

HUMBER'S RECORD OF MODERN ENGINEERING. THIRD SERIES. Imp. 4to, with 40 Double Plates, Photographic Portrait of J. R. M'Clean, late Pres. Inst. C.E., and copious descriptive Letterpress, Specifications, &c., £3 3s. half-morocco.

List of the Plates and Diagrams.

MAIN DRAINAGE, METROPOLIS.—*North Side.*—Map showing Interception of Sewers; Middle Level Sewer (2 plates); Outfall Sewer, Bridge over River Lea (3 plates); Outfall Sewer, Bridge over Marsh Lane, North Woolwich Railway, and Bow and Barking Railway Junction; Outfall Sewer, Bridge over Bow and Barking Railway (3 plates); Outfall Sewer, Bridge over East London Waterworks' Feeder (2 plates); Outfall Sewer, Reservoir (2 plates); Outfall Sewer, Tumbling Bay and Outlet; Outfall Sewer, Penstocks. *South Side.*—Outfall Sewer, Bermondsey Branch (2 plates); Outfall Sewer, Reservoir and Outlet (4 plates); Outfall Sewer, Filth Hoist; Sections of Sewers (North and South Sides). THAMES EMBANKMENT.—Section of River Wall; Steamboat Pier, Westminster (2 plates); Landing Stairs between Charing Cross and Waterloo Bridges; York Gate (2 plates); Overflow and Outlet at Savoy Street Sewer (3 plates); Steamboat Pier, Waterloo Bridge (3 plates); Junction of Sewers, Plans and Sections; Gullies, Plans and Sections; Rolling Stock; Granite and Iron Forts.

"The drawings have a constantly increasing value, and whoever desires to possess clear representations of the two great works carried out by our Metropolitan Board will obtain Mr. Humber's volume."—*Engineer.*

HUMBER'S RECORD OF MODERN ENGINEERING. FOURTH SERIES. Imp. 4to, with 36 Double Plates, Photographic Portrait of John Fowler, late Pres. Inst. C.E., and copious descriptive Letterpress, Specifications, &c., £3 3s. half-morocco.

List of the Plates and Diagrams.

Abbey Mills Pumping Station, Main Drainage, Metropolis (4 plates); Barrow Docks (5 plates); Manquis Viaduct, Santiago and Valparaiso Railway (2 plates); Adam's Locomotive, St. Helen's Canal Railway (2 plates); Cannon Street Station Roof, Charing Cross Railway (3 plates); Road Bridge over the River Moka (2 plates); Telegraphic Apparatus for Mesopotamia; Viaduct over the River Wye, Midland Railway (3 plates); St. Germans Viaduct, Cornwall Railway (2 plates); Wrought-Iron Cylinder for Diving Bell; Millwall Docks (6 plates); Milroy's Patent Excavator; Metropolitan District Railway (6 plates); Harbours, Ports, and Breakwaters (3 plates).

"We gladly welcome another year's issue of this valuable publication from the able pen of Mr. Humber. The accuracy and general excellence of this work are well known, while its usefulness in giving the measurements and details of some of the latest examples of engineering, as carried out by the most eminent men in the profession, cannot be too highly prized."—*Artisan.*

CIVIL ENGINEERING, SURVEYING, etc. 9

MR. HUMBER'S ENGINEERING BOOKS—continued.

Strains, Calculation of.
A HANDY BOOK FOR THE CALCULATION OF STRAINS IN GIRDERS AND SIMILAR STRUCTURES, AND THEIR STRENGTH. Consisting of Formulæ and Corresponding Diagrams, with numerous details for Practical Application, &c. By WILLIAM HUMBER, A-M.Inst.C.E., &c. Fourth Edition. Crown 8vo, nearly 100 Woodcuts and 3 Plates, 7s. 6d. cloth.
"The formulæ are neatly expressed, and the diagrams good."—*Athenæum.*
"We heartily commend this really *handy* book to our engineer and architect readers."—*English Mechanic.*

Barlow's Strength of Materials, enlarged by Humber
A TREATISE ON THE STRENGTH OF MATERIALS; with Rules for Application in Architecture, the Construction of Suspension Bridges, Railways, &c. By PETER BARLOW, F.R.S. A New Edition, revised by his Sons, P. W. BARLOW, F.R.S., and W. H. BARLOW, F.R.S.; to which are added, Experiments by HODGKINSON, FAIRBAIRN, and KIRKALDY; and Formulæ for Calculating Girders, &c. Arranged and Edited by W. HUMBER, A-M.Inst.C.E. Demy 8vo, 400 pp., with 19 large Plates and numerous Woodcuts, 18s. cloth.
"Valuable alike to the student, tyro, and the experienced practitioner, it will always rank in future, as it has hitherto done, as the standard treatise on that particular subject."—*Engineer.*
"There is no greater authority than Barlow."—*Building News.*
"As a scientific work of the first class, it deserves a foremost place on the bookshelves of every civil engineer and practical mechanic."—*English Mechanic.*

Trigonometrical Surveying.
AN OUTLINE OF THE METHOD OF CONDUCTING A TRIGONOMETRICAL SURVEY, for the Formation of Geographical and Topographical Maps and Plans, Military Reconnaissance, Levelling, &c., with Useful Problems, Formulæ, and Tables. By Lieut.-General FROME, R.E. Fourth Edition, Revised and partly Re-written by Major General Sir CHARLES WARREN, G.C.M.G., R.E. With 19 Plates and 115 Woodcuts, royal 8vo, 16s. cloth.
"The simple fact that a fourth edition has been called for is the best testimony to its merits. No words of praise from us can strengthen the position so well and so steadily maintained by this work. Sir Charles Warren has revised the entire work, and made such additions as were necessary to bring every portion of the contents up to the present date."—*Broad Arrow.*

Field Fortification.
A TREATISE ON FIELD FORTIFICATION, THE ATTACK OF FORTRESSES, MILITARY MINING, AND RECONNOITRING. By Colonel I. S. MACAULAY, late Professor of Fortification in the R.M.A., Woolwich. Sixth Edition, crown 8vo, cloth, with separate Atlas of 12 P.ates, 12s.

Oblique Bridges.
A PRACTICAL AND THEORETICAL ESSAY ON OBLIQUE BRIDGES. With 13 large Plates. By the late GEORGE WATSON BUCK, M.I.C.E. Third Edition, revised by his Son, J. H. WATSON BUCK, M.I.C.E.; and with the addition of Description to Diagrams for Facilitating the Construction of Oblique Bridges, by W. H. BARLOW, M.I.C.E. Royal 8vo, 12s. cloth.
"The standard text-book for all engineers regarding skew arches is Mr. Buck's treatise, and it would be impossible to consult a better."—*Engineer.*
"Mr. Buck's treatise is recognised as a standard text-book, and his treatment has divested the subject of many of the intricacies supposed to belong to it. As a guide to the engineer and architect, on a confessedly difficult subject, Mr. Buck's work is unsurpassed."—*Building News.*

Water Storage, Conveyance and Utilisation.
WATER ENGINEERING: A Practical Treatise on the Measurement, Storage, Conveyance and Utilisation of Water for the Supply of Towns, for Mill Power, and for other Purposes. By CHARLES SLAGG, Water and Drainage Engineer, A.M.Inst.C.E., Author of "Sanitary Work in the Smaller Towns, and in Villages," &c. With numerous Illusts. Cr. 8vo, 7s. 6d. cloth.
"As a small practical treatise on the water supply of towns, and on some applications of water-power, the work is in many respects excellent."—*Engineering.*
"The author has collated the results deduced from the experiments of the most eminent authorities, and has presented them in a compact and practical form, accompanied by very clear and detailed explanations. . . . The application of water as a motive power is treated very carefully and exhaustively."—*Builder.*
"For anyone who desires to begin the study of hydraulics with a consideration of the practica' applications of the science there is no better guide."—*Architect.*

Statics, Graphic and Analytic.

GRAPHIC AND ANALYTIC STATICS, in their Practical Application to the Treatment of Stresses in Roofs, Solid Girders, Lattice, Bowstring and Suspension Bridges, Braced Iron Arches and Piers, and other Frameworks. By R. HUDSON GRAHAM, C.E. Containing Diagrams and Plates to Scale. With numerous Examples, many taken from existing Structures. Specially arranged for Class-work in Colleges and Universities. Second Edition, Revised and Enlarged. 8vo, 16s. cloth.

"Mr. Graham's book will find a place wherever graphic and analytic statics are used or studied."
—*Engineer*.
"The work is excellent from a practical point of view, and has evidently been prepared with much care. The directions for working are ample, and are illustrated by an abundance of well-selected examples. It is an excellent text-book for the practical draughtsman."—*Athenæum*.

Student's Text-Book on Surveying.

PRACTICAL SURVEYING: A Text-Book for Students preparing for Examination or for Survey-work in the Colonies. By GEORGE W. USILL, A.M.I.C.E., Author of "The Statistics of the Water Supply of Great Britain." With Four Lithographic Plates and upwards of 330 Illustrations. Second Edition, Revised. Crown 8vo, 7s. 6d. cloth. [*Just published*.

"The best forms of instruments are described as to their construction, uses and modes of employment, and there are innumerable hints on work and equipment such as the author, in his experience as surveyor, draughtsman and teacher, has found necessary, and which the student in his inexperience will find most serviceab'e."—*Engineer*.
"The latest treatise in the English language on surveying, and we have no hesitation in saying that the student will find it a be.ter guide than any of its predecessors. . . . . De erves to be recognised as the first book which should be put in the hands of a pupil of Civil Engineering, and every gentleman of education who sets out for the Colonies would find it well to have a copy."—*Architect*.
"A very useful, practical handbook on field practice. Clear, accurate and not too condensed."—*Journal of Education*.

Survey Practice.

AID TO SURVEY PRACTICE, for Reference in Surveying, Levelling, and Setting-out; and in Route Surveys of Travellers by Land and Sea. With Tables, Illustrations, and Records. By LOWIS D'A. JACKSON, A.M.I.C.E., Author of "Hydraulic Manual," "Modern Metrology," &c. Second Edition, Enlarged. Large crown 8vo, 12s. 6d. cloth.

"Mr. Jackson has produced a valuable *vade-mecum* for the surveyor. We can recommend this book as containing an admirable supplement to the teaching of the accomplished surveyor."—*Athenæum*.
"As a text-book we should advise all surveyors to place it in their libraries, and study well the matured instructions afforded in its pages."—*Colliery Guardian*.
"The author brings to his work a fortunate union of theory and practical experience which, aided by a clear and lucid style of writing, renders the book a very useful one."—*Builder*.

Surveying, Land and Marine.

LAND AND MARINE SURVEYING, in Reference to the Preparation of Plans for Roads and Railways; Canals, Rivers, Towns' Water Supplies; Docks and Harbours. With Description and Use of Surveying Instruments. By W. D. HASKOLL, C.E., Author of "Bridge and Viaduct Construction," &c. Second Edition, Revised, with Additions. Large cr. 8vo, 9s. cl.

"This book must prove of great value to the student. We have no hesitation in recommending it, feeling assured that it will more than repay a careful study."—*Mechanical World*.
"A most u e'ul and well arranged book for the aid of a student. We can strongly recommend it as a carefully-written and valuable text-book. It enjoys a well-deserved repute among surveyors."
—*Builder*.
"This volume cannot fail to prove of the utmost practical utility. It may be safely recommended to all students who aspire to become clean and expert surveyors."—*Mining Journal*.

Tunnelling.

PRACTICAL TUNNELLING. Explaining in detail the Setting-out of the works, Shaft-sinking and Heading-driving, Ranging the Lines and Levelling underground, Sub-Excavating, Timbering, and the Construction of the Brickwork of Tunnels, with the amount of Labour required for, and the Cost of, the various portions of the work. By FREDERICK W. SIMMS, F.G.S., M.Inst.C.E. Third Edition, Revised and Extended by D. KINNEAR CLARK, M.Inst.C.E. Imperial 8vo, with 21 Folding Plates and numerous Wood Engravings, 30s. cloth.

"The estimation in which Mr. Simms's book on tunnelling has been held for over thirty years cannot be more truly expressed than in the words of the late Prof. Rankine:—' The best source of information on the subject of tunnels is Mr. F. W. Simms's work on Practical Tunnelling.'"—*Architect*.
"It has been regarded from the first as a text book of the subject. . . . Mr. Clarke has added immensely to the value of the book."—*Engineer*.

CIVIL ENGINEERING, SURVEYING, etc. 11

Levelling.

A TREATISE ON THE PRINCIPLES AND PRACTICE OF LEVELLING. Showing its Application to purposes of Railway and Civil Engineering, in the Construction of Roads; with Mr. TELFORD's Rules for the same. By FREDERICK W. SIMMS, F.G.S., M.Inst.C.E. Seventh Edition, with the addition of LAW's Practical Examples for Setting-out Railway Curves, and TRAUTWINE's Field Practice of Laying-out Circular Curves. With 7 Plates and numerous Woodcuts, 8vo, 8s. 6d. cloth. *₊* TRAUTWINE on Curves may be had separate, 5s.

"The text-book on levelling in most of our engineering schools and colleges."—*Engineer*.
"The publishers have rendered a substantial service to the profession, especially to the younger members, by bringing out the present edition of Mr. Simms's useful work."—*Engineering*.

Heat, Expansion by.

EXPANSION OF STRUCTURES BY HEAT. By JOHN KEILY, C.E., late of the Indian Public Works and Victorian Railway Departments. Crown 8vo, 3s. 6d. cloth.

SUMMARY OF CONTENTS.
Section I. FORMULAS AND DATA.
Section II. METAL BARS.
Section III. SIMPLE FRAMES.
Section IV. COMPLEX FRAMES AND PLATES.
Section V. THERMAL CONDUCTIVITY.
Section VI. MECHANICAL FORCE OF HEAT.
Section VII. WORK OF EXPANSION AND CONTRACTION.
Section VIII. SUSPENSION BRIDGES.
Section IX. MASONRY STRUCTURES.

"The aim the author has set before him, viz., to show the effects of heat upon metallic and other structures, is a laudable one, for this is a branch of physics upon which the engineer or architect can find but little reliable and comprehensive data in books."—*Builder*.
"Whoever is concerned to know the effect of changes of temperature on such structures as suspension bridges and the like, could not do better than consult Mr. Keily's valuable and handy exposition of the geometrical principles involved in these changes."—*Scotsman*.

Practical Mathematics.

MATHEMATICS FOR PRACTICAL MEN: Being a Commonplace Book of Pure and Mixed Mathematics. Designed chiefly for the use of Civil Engineers, Architects and Surveyors. By OLINTHUS GREGORY, LL.D., F.R.A.S., enlarged by HENRY LAW, C.E. 4th Edition, carefully Revised by J. R. YOUNG, formerly Professor of Mathematics, Belfast College. With 13 Plates, 8vo, £1 1s. cloth.

"The engineer or architect will here find ready to his hand rules for solving nearly every mathematical difficulty that may arise in his practice. The rules are in all cases explained by means of examples, in which every step of the process is clearly worked out."—*Builder*.
"One of the most serviceable books for practical mechanics. . . . It is an instructive book for the student, and a text-book for him who, having once mastered the subjects it treats of, needs occasionally to refresh his memory upon them."—*Building News*.

Hydraulic Tables.

HYDRAULIC TABLES, CO-EFFICIENTS, and FORMULÆ for finding the Discharge of Water from Orifices, Notches, Weirs, Pipes, and Rivers. With New Formulæ, Tables, and General Information on Rainfall, Catchment-Basins, Drainage, Sewerage, Water Supply for Towns and Mill Power. By JOHN NEVILLE, Civil Engineer, M.R.I.A. Third Ed., carefully Revised, with considerable Additions. Numerous Illusts. Cr. 8vo, 14s. cloth.

"Alike valuable to students and engineers in practice; its study will prevent the annoyance of avoidable failures, and assist them to select the readiest means of successfully carrying out any given work connected with hydraulic engineering."—*Mining Journal*.
"It is, of all English books on the subject, the one nearest to completeness. . . . From the good arrangement of the matter, the clear explanations, and abundance of formulæ, the carefully calculated tables, and, above all, the thorough acquaintance with both theory and construction, which is displayed from first to last, the book will be found to be an acquisition."—*Architect*.

Hydraulics.

HYDRAULIC MANUAL. Consisting of Working Tables and Explanatory Text. Intended as a Guide in Hydraulic Calculations and Field Operations. By LOWIS D'A. JACKSON, Author of "Aid to Survey Practice," "Modern Metrology," &c. Fourth Edition, Enlarged. Large cr. 8vo, 16s. cl.

"The author has had a wide experience in hydraulic engineering and has been a careful observer of the facts which have come under his notice, and from the great mass of material at his command he has constructed a manual which may be accepted as a trustworthy guide to this branch of the engineer's profession. We can heartily recommend this volume to all who desire to be acquainted with the latest development of this important subject."—*Engineering*.
"The standard-work in this department of mechnnics."—*Scotsman*.
"The most useful feature of this work is its freedom from what is superannuated, and its thorough adoption of recent experiments; the text is, in fact, in great part a short account of the great modern experiments."—*Nature*.

Drainage.

ON THE DRAINAGE OF LANDS, TOWNS AND BUILD-INGS. By G. D. DEMPSEY, C.E., Author of "The Practical Railway Engineer," &c. Revised, with large Additions on RECENT PRACTICE IN DRAINAGE ENGINEERING, by D. KINNEAR CLARK, M.Inst.C.E. Author of "Tramways: Their Construction and Working," "A Manual of Rules, Tables, and Data for Mechanical Engineers," &c. &c. Crown 8vo, 7s. 6d. cloth.
[Just published.
"The new matter added to Mr. Dempsey's excellent work is characterised by the comprehensive grasp and accuracy of detail for which the name of Mr. D. K. Clark is a sufficient voucher."—*Athenæum.*
"As a work on recent practice in drainage engineering, the book is to be commended to all who are making that branch of engineering science their special study."—*Iron.*
"A comprehensive manual on drainage engineering, and a useful introduction to the student."—*Building News.*

Tramways and their Working.

TRAMWAYS: THEIR CONSTRUCTION AND WORKING. Embracing a Comprehensive History of the System; with an exhaustive Analysis of the various Modes of Traction, including Horse-Power, Steam, Heated Water, and Compressed Air; a Description of the Varieties of Rolling Stock; and ample Details of Cost and Working Expenses: the Progress recently made in Tramway Construction, &c. &c. By D. KINNEAR CLARK, M.Inst.C.E. With over 200 Wood Engravings, and 13 Folding Plates. Two Vols., large crown 8vo, 30s. cloth.
"All interested in tramways must refer to it, as all railway engineers have turned to the author's work 'Railway Machinery.'"—*Engineer.*
"An exhaustive and practical work on tramways, in which the history of this kind of locomotion, and a description and cost of the various modes of laying tramways, are to be found."—*Building News.*
"The best form of rails, the best mode of construction, and the best mechanical appliances are so fairly indicated in the work under review, that any engineer about to construct a tramway will be enabled at once to obtain the practical information which will be of most service to him."—*Athenæum.*

Oblique Arches.

A PRACTICAL TREATISE ON THE CONSTRUCTION OF OBLIQUE ARCHES. By JOHN HART. Third Edition, with Plates. Imperial 8vo, 8s. cloth.

Curves, Tables for Setting-out.

TABLES OF TANGENTIAL ANGLES AND MULTIPLES *for Setting-out Curves from* 5 *to* 200 *Radius.* By ALEXANDER BEAZELEY, M.Inst.C.E. Third Edition. Printed on 48 Cards, and sold in a cloth box, waistcoat-pocket size, 3s. 6d.
"Each table is printed on a small card, which, being placed on the theodolite, leaves the hands free to manipulate the instrument—no small advantage as regards the rapidity of work."—*Engineer.*
"Very handy; a man may know that all his day's work must fall on two of these cards, which he puts into his own card-case, and leaves the rest behind."—*Athenæum.*

Earthwork.

EARTHWORK TABLES. Showing the Contents in Cubic Yards of Embankments, Cuttings, &c., of Heights or Depths up to an average of 80 feet. By JOSEPH BROADBENT, C.E., and FRANCIS CAMPIN, C.E. Crown 8vo, 5s. cloth.
"The way in which accuracy is attained, by a simple division of each cross section into three elements, two in which are constant and one variable, is ingenious."—*Athenæum.*

Tunnel Shafts.

THE CONSTRUCTION OF LARGE TUNNEL SHAFTS: A *Practical and Theoretical Essay.* By J. H. WATSON BUCK, M.Inst.C.E., Resident Engineer, London and North-Western Railway. Illustrated with Folding Plates, royal 8vo, 12s. cloth.
"Many of the methods given are of extreme practical value to the mason; and the observations on the form of arch, the rules for ordering the stone, and the construction of the templates will be found of considerable use. We commend the book to the engineering profession."—*Building News.*
"Will be regarded by civil engineers as of the utmost value, and calculated to save much time and obviate many mistakes."—*Colliery Guardian.*

Girders, Strength of.

GRAPHIC TABLE FOR FACILITATING THE COMPUTATION OF THE WEIGHTS OF WROUGHT IRON AND STEEL GIRDERS, etc., for Parliamentary and other Estimates. By J. H. WATSON BUCK, M.Inst.C.E. On a Sheet, 2s. 6d.

CIVIL ENGINEERING, SURVEYING, etc. 13

River Engineering.
RIVER BARS: The Causes of their Formation, and their Treatment by "Induced Tidal Scour;" with a Description of the Successful Reduction by this Method of the Bar at Dublin. By I. J. MANN, Assist. Eng. to the Dublin Port and Docks Board. Royal 8vo, 7s. 6d. cloth.

"We recommend all interested in harbour works—and, indeed, those concerned in the improvements of rivers generally—to read Mr. Mann's interesting work on the treatment of river bars."—*Engineer*.

Trusses.
TRUSSES OF WOOD AND IRON. Practical Applications of Science in Determining the Stresses, Breaking Weights, Safe Loads, Scantlings, and Details of Construction, with Complete Working Drawings. By WILLIAM GRIFFITHS, Surveyor, Assistant Master, Tranmere School of Science and Art. Oblong 8vo, 4s. cloth.

"This handy little book enters so minutely into every detail connected with the construction of roof trusses, that no student need be ignorant of these matters."—*Practical Engineer*.

Railway Working.
SAFE RAILWAY WORKING. A Treatise on Railway Accidents: Their Cause and Prevention; with a Description of Modern Appliances and Systems. By CLEMENT E. STRETTON, C.E., Vice-President and Consulting Engineer, Amalgamated Society of Railway Servants. With Illustrations and Coloured Plates. Second Edition, Enlarged. Crown 8vo, 3s. 6d. cloth. [*Just published*.

"A book for the engineer, the directors, the managers; and, in short, all who wish for information on railway matters will find a perfect encyclopædia in 'Safe Railway Working.'"—*Railway Review*.

"We commend the remarks on railway signalling to all railway managers, especially where a uniform code and practice is advocated."—*Herepath's Railway Journal*.

"The author may be congratulated on having collected, in a very convenient form, much valuable information on the principal questions affecting the safe working of railways."—*Railway Engineer*.

Field-Book for Engineers.
THE ENGINEER'S, MINING SURVEYOR'S, AND CONTRACTOR'S FIELD-BOOK. Consisting of a Series of Tables, with Rules, Explanations of Systems, and use of Theodolite for Traverse Surveying and Plotting the Work with minute accuracy by means of Straight Edge and Set Square only; Levelling with the Theodolite, Casting-out and Reducing Levels to Datum, and Plotting Sections in the ordinary manner; setting-out Curves with the Theodolite by Tangential Angles and Multiples, with Right and Left-hand Readings of the Instrument; Setting-out Curves without Theodolite, on the System of Tangential Angles by sets of Tangents and Offsets; and Earthwork Tables to 80 feet deep, calculated for every 6 inches in depth. By W. DAVIS HASKOLL, C.E. With numerous Woodcuts. Fourth Edition, Enlarged. Crown 8vo, 12s. cloth.

"The book is very handy; the separate tables of sines and tangents to every minute will make t useful for many other purposes, the genuine traverse tables existing all the same."—*Athenæum*.

"Every person engaged in engineering field operations will estimate the importance of such a work and the amount of valuable time which will be saved by reference to a set of reliable tables prepared with the accuracy and fulness of those given in this volume."—*Railway News*.

Earthwork, Measurement of.
A MANUAL ON EARTHWORK. By ALEX. J. S. GRAHAM, C.E. With numerous Diagrams. Second Edition. 18mo, 2s. 6d. cloth.

"A great amount of practical information, very admirably arranged, and available for rough estimates, as well as for the more exact calculations required in the engineer's and contractor's offices."—*Artisan*.

Strains in Ironwork.
THE STRAINS ON STRUCTURES OF IRONWORK; with Practical Remarks on Iron Construction. By F. W. SHEILDS, M.Inst.C.E. Second Edition, with 5 Plates. Royal 8vo, 5s. cloth.

"The student cannot find a better little book on this subject."—*Engineer*.

Cast Iron and other Metals, Strength of.
A PRACTICAL ESSAY ON THE STRENGTH OF CAST IRON AND OTHER METALS. By THOMAS TREDGOLD, C.E. Fifth Edition, including HODGKINSON's Experimental Researches. 8vo, 12s. cloth.

ARCHITECTURE, BUILDING, etc.

Construction.
THE SCIENCE OF BUILDING: An Elementary Treatise on the Principles of Construction. By E. WYNDHAM TARN, M.A., Architect. Third Edition, Revised and Enlarged, with 59 Engravings. Fcap. 8vo, 4s. cloth. [*Just published.*
"A very valuable book, which we strongly recommend to all students."—*Builder.*
"No architectural student should be without this handbook of constructional knowledge."—*Architect.*

Villa Architecture.
A HANDY BOOK OF VILLA ARCHITECTURE: Being a Series of Designs for Villa Residences in various Styles. With Outline Specifications and Estimates. By C. WICKES, Architect, Author of "The Spires and Towers of England," &c. 61 Plates, 4to, £1 11s. 6d. half-morocco, gilt edges.
"The whole of the designs bear evidence of their being the work of an artistic architect, and they will prove very valuable and suggestive."—*Building News.*

Text-Book for Architects.
THE ARCHITECT'S GUIDE: Being a Text-Book of Useful Information for Architects, Engineers, Surveyors, Contractors, Clerks of Works, &c. &c. By FREDERICK ROGERS, Architect, Author of "Specifications for Practical Architecture," &c. Second Edition, Revised and Enlarged. With numerous Illustrations. Crown 8vo, 6s. cloth.
"As a text-book of useful information for architects, engineers, surveyors, &c., it would be hard to find a handier or more complete little volume."—*Standard.*
"A young architect could hardly have a better guide-book."—*Timber Trades Journal.*

Taylor and Cresy's Rome.
THE ARCHITECTURAL ANTIQUITIES OF ROME. By the late G. L. TAYLOR, Esq., F.R.I.B.A., and EDWARD CRESY, Esq. New Edition, thoroughly Revised by the Rev. ALEXANDER TAYLOR, M.A. (son of the late G. L. Taylor, Esq.), Fellow of Queen's College, Oxford, and Chaplain of Gray's Inn. Large folio, with 130 Plates, half-bound, £3 3s.
N.B.—*This is the only book which gives on a large scale, and with the precision of architectural measurement, the principal Monuments of Ancient Rome in plan, elevation, and detail.*
Taylor and Cresy's work has from its first publication been ranked among those professional books which cannot be bettered. . . . It would be difficult to find examples of drawings, even among those of the most painstaking students of Gothic, more thoroughly worked out than are the one hundred and thirty plates in this volume."—*Architect.*

Architectural Drawing.
PRACTICAL RULES ON DRAWING, for the Operative Builder and Young Student in Architecture. By GEORGE PYNE. With 14 Plates, 4to, 7s. 6d. boards.

Sir Wm. Chambers's Treatise on Civil Architecture.
THE DECORATIVE PART OF CIVIL ARCHITECTURE. By Sir WILLIAM CHAMBERS, F.R.S. With Portrait, Illustrations, Notes, and an Examination of Grecian Architecture, by JOSEPH GWILT, F.S.A. Revised and Edited by W. H. LEEDS, with a Memoir of the Author. 66 Plates, 4to, 21s. cloth.

House Building and Repairing.
THE HOUSE-OWNER'S ESTIMATOR; or, What will it Cost to Build, Alter, or Repair? A Price Book adapted to the Use of Unprofessional People, as well as for the Architectural Surveyor and Builder. By JAMES D. SIMON, A.R.I.B.A. Edited and Revised by FRANCIS T. W. MILLER, A.R.I.B.A. With numerous Illustrations. Fourth Edition, Revised. Crown 8vo, 3s. 6d. cloth.
"In two years it will repay its cost a hundred times over."—*Field.*
"A very handy book."—*English Mechanic.*

Cottages and Villas.
COUNTRY AND SUBURBAN COTTAGES AND VILLAS: How to Plan and Build Them. Containing 33 Plates, with Introduction, General Explanations, and Description of each Plate. By JAMES W. BOGUE, Architect, Author of "Domestic Architecture," &c. 4to, 10s. 6d. cloth.
[*Just published.*

ARCHITECTURE, BUILDING, etc. 15

The New Builder's Price Book, 1891.

LOCKWOOD'S BUILDER'S PRICE BOOK FOR 1891. A Comprehensive Handbook of the Latest Prices and Data for Builders, Architects, Engineers and Contractors. Re-constructed, Re-written and Greatly Enlarged. By FRANCIS T. W. MILLER. 640 closely-printed pages, crown 8vo, 4s. cloth. [*Just published.*
"This book is a very useful one, and should find a place in every English office connected with the building and engineering professions."—*Industries.*
"This Price Book has been set up in new type. . . . Advantage has been taken of the transformation to add much additional information, and the volume is now an excellent book of reference."—*Architect.*
"In its new and revised form this Price Book is what a work of this kind should be—comprehensive, reliable, well arranged, legible and well b und."—*British Architect.*
"A work of established repuiation."—*Athenæum.*
"This very useful handbook is well written, exceedingly clear in its explanations and great care has evidently been taken to ensure accuracy."—*Morning Advertiser.*

Designing, Measuring, and Valuing.

THE STUDENT'S GUIDE to the PRACTICE of MEASURING AND VALUING ARTIFICERS' WORKS. Containing Directions for taking Dimensions, Abstracting the same, and bringing the Quantities into Bill, with Tables of Constants for Valuation of Labour, and for the Calculation of Areas and Solidities. Originally edited by EDWARD DOBSON, Architect. With Additions on Mensuration and Construction, and a New Chapter on Dilapidations, Repairs, and Contracts, by E. WYNDHAM TARN, M.A. Sixth Edition, including a Complete Form of a Bill of Quantities. With 8 Plates and 63 Woodcuts. Crown 8vo, 7s. 6d. cloth.
"Well fulfils the promise of its title-page, and we can thoroughly recommend it to the class for whose use it has been compiled. Mr. Tarn's additions and revisions have much increased the usefulness of the work, and have especially augmented its value to students."—*Engineering.*
"This edition will be found the most complete treatise on the principles of measuring and valuing artificers' work that has yet been published."—*Building News.*

Pocket Estimator and Technical Guide.

THE POCKET TECHNICAL GUIDE, MEASURER AND ESTIMATOR FOR BUILDERS AND SURVEYORS. Containing Technical Directions for Measuring Work in all the Building Trades, Complete Specifications for Houses, Roads, and Drains, and an easy Method of Estimating the parts of a Building collectively. By A. C. BEATON, Author of "Quantities and Measurements," &c. Fifth Edition. With 53 Woodcuts, waistcoat-pocket size, 1s. 6d. gilt edges.
"No builder, architect, surveyor, or valuer should be without his 'Beaton.'"—*Building News.*
"Contains an extraordinary amount of information in daily requisition in measuring and estimating. Its presence in the pocket will save valuable time and trouble."—*Building World.*

Donaldson on Specifications.

THE HANDBOOK OF SPECIFICATIONS; or, Practical Guide to the Architect, Engineer, Surveyor, and Builder, in drawing up Specifications and Contracts for Works and Constructions. Illustrated by Precedents of Buildings actually executed by eminent Architects and Engineers. By Professor T. L. DONALDSON, P.R.I.B.A., &c. New Edition, in One large Vol., 8vo, with upwards of 1,000 pages of Text, and 33 Plates, £1 11s. 6d. cloth.
"In this work forty-four specifications of executed works are given, including the specifications for parts of the new Houses of Parliament, by Sir Charles Barry, and for the new Royal Exchange, by Mr. Tite, M.P. The latter, in particular, is a very complete and remarkable document. It embodies, to a great extent, as Mr. Donaldson mentions, 'the bill of quantities with the description of the works.' . . . It is valuable as a record, and more valuable still as a book of precedents. . . . Suffice it to say that Donaldson's 'Handbook of Specifications' must be bought by all architects."—*Builder.*

Bartholomew and Rogers' Specifications.

SPECIFICATIONS FOR PRACTICAL ARCHITECTURE. A Guide to the Architect, Engineer, Surveyor, and Builder. With an Essay on the Structure and Science of Modern Buildings. Upon the Basis of the Work by ALFRED BARTHOLOMEW, thoroughly Revised, Corrected, and greatly added to by FREDERICK ROGERS, Architect. Second Edition, Revised, with Additions. With numerous Illustrations, medium 8vo, 15s. cloth.
"The collection of specifications prepared by Mr. Rogers on the basis of Bartholomew's work is too well known to need any recommendation from us. It is one of the books with which every young architect must be equipped; for time has shown that the specifications cannot be set aside through any defect in them."—*Architect.*

Building; Civil and Ecclesiastical.

A BOOK ON BUILDING, Civil and Ecclesiastical, including Church Restoration; with the Theory of Domes and the Great Pyramid, &c. By Sir EDMUND BECKETT, Bart., LL.D., F.R.A.S., Author of "Clocks and Watches, and Bells," &c. Second Edition, Enlarged. Fcap. 8vo, 5s. cloth.
"A book which is always amusing and nearly always instructive. The style throughout is in the highest degree condensed and epigrammatic."—*Times.*

Ventilation of Buildings.

VENTILATION. *A Text Book to the Practice of the Art of Ventilating Buildings.* With a Chapter upon Air Testing. By W. P. BUCHAN, R.P., Sanitary and Ventilating Engineer, Author of "Plumbing," &c. With 170 Illustrations. 12mo, 4s. cloth boards. [*Just published.*

The Art of Plumbing.

PLUMBING. *A Text Book to the Practice of the Art or Craft of the Plumber,* with Supplementary Chapters on House Drainage, embodying the latest Improvements. By WILLIAM PATON BUCHAN, R.P., Sanitary Engineer and Practical Plumber. Fifth Edition, Enlarged to 370 pages, and 380 Illustrations. 12mo, 4s. cloth boards.
"A text book which may be safely put in the hands of every young plumber, and which will also be found useful by architects and medical professors."—*Builder.*
"A valuable text book, and the only treatise which can be regarded as a really reliable manual of the plumber's art."—*Building News.*

Geometry for the Architect, Engineer, etc.

PRACTICAL GEOMETRY, *for the Architect, Engineer and Mechanic.* Giving Rules for the Delineation and Application of various Geometrical Lines, Figures and Curves. By E. W. TARN, M.A., Architect, Author of "The Science of Building," &c. Second Edition. With 172 Illustrations, demy 8vo, 9s. cloth.
"No book with the same objects in view has ever been published in which the clearness of the rules laid down and the illustrative diagrams have been so satisfactory."—*Scotsman.*

The Science of Geometry.

THE GEOMETRY OF COMPASSES; *or, Problems Resolved by the mere Description of Circles, and the use of Coloured Diagrams and Symbols.* By OLIVER BYRNE. Coloured Plates. Crown 8vo, 3s. 6d. cloth.
"The treatise is a good one, and remarkable—like all Mr. Byrne's contributions to the science of geometry—for the lucid character of its teaching."—*Building News.*

DECORATIVE ARTS, etc.

Woods and Marbles (Imitation of).

SCHOOL OF PAINTING FOR THE IMITATION OF WOODS AND MARBLES, as Taught and Practised by A. R. VAN DER BURG and P. VAN DER BURG, Directors of the Rotterdam Painting Institution. Royal folio, 18¾ by 12¼ in., Illustrated with 24 full-size Coloured Plates; also 12 plain Plates, comprising 154 Figures. Second and Cheaper Edition. Price £1 11s. 6d.

List of Plates.

1. Various Tools required for Wood Painting—2, 3. Walnut: Preliminary Stages of Graining and Finished Specimen—4. Tools used for Marble Painting and Method of Manipulation—5, 6. St. Remi Marble: Earlier Operations and Finished Specimen—7. Methods of Sketching different Grains, Knots, &c.—8, 9. Ash: Preliminary Stages and Finished Specimen—10. Methods of Sketching Marble Grains—11, 12. Breche Marble: Preliminary Stages of Working and Finished Specimen—13. Maple: Methods of Producing the different Grains—14, 15. Bird's-eye Maple: Preliminary Stages and Finished Specimen—16. Methods of Sketching the different Species of White Marble—17, 18. White Marble: Preliminary Stages of Process and Finished Specimen—19. Mahogany: Specimens of various Grains and Methods of Manipulation—20, 21. Mahogany: Earlier Stages and Finished Specimen—22, 23, 24. Sienna Marble: Varieties of Grain, Preliminary Stages and Finished Specimen—25, 26, 27. Juniper Wood: Methods of producing Grain, &c.: Preliminary Stages and Finished Specimen—28, 29, 30. Vert de Mer Marble: Varieties of Grain and Methods of Working Unfinished and Finished Specimens—31, 32, 33. Oak: Varieties of Grain, Tools Employed, and Methods of Manipulation, Preliminary Stages and Finished Specimen—34, 35, 36. Waulsort Marble: Varieties of Grain, Unfinished and Finished Specimens.

* OPINIONS OF THE PRESS.

"Those who desire to attain skill in the art of painting woods and marbles will find advantage in consulting this book. . . . Some of the Working Men's Clubs should give their young men the opportunity to study it."—*Builder.*
"A comprehensive guide to the art. The explanations of the processes, the manipulation and management of the colours, and the beautifully executed plates will not be the least valuable to the student who aims at making his work a faithful transcript of nature."—*Building News.*

House Decoration.
ELEMENTARY DECORATION. A Guide to the Simpler Forms of Everyday Art, as applied to the Interior and Exterior Decoration of Dwelling Houses, &c. By JAMES W. FACEY, Jun. With 68 Cuts. 12mo, 2s. cloth limp.

PRACTICAL HOUSE DECORATION: A Guide to the Art of Ornamental Painting, the Arrangement of Colours in Apartments, and the principles of Decorative Design. With some Remarks upon the Nature and Properties of Pigments. By JAMES WILLIAM FACEY, Author of "Elementary Decoration," &c. With numerous Illustrations. 12mo, 2s. 6d. cloth limp.

N.B.—The above Two Works together in One Vol., strongly half-bound, 5s.

Colour.
A GRAMMAR OF COLOURING. Applied to Decorative Painting and the Arts. By GEORGE FIELD. New Edition, Revised, Enlarged, and adapted to the use of the Ornamental Painter and Designer. By ELLIS A. DAVIDSON. With New Coloured Diagrams and Engravings. 12mo, 3s. 6d. cloth boards.

"The book is a most useful *resume* of the properties of pigments."—*Builder.*

House Painting, Graining, etc.
HOUSE PAINTING, GRAINING, MARBLING, AND SIGN WRITING, A Practical Manual of. By ELLIS A. DAVIDSON. Fifth Edition. With Coloured Plates and Wood Engravings. 12mo, 6s. cloth boards.

"A mass of information, of use to the amateur and of value to the practical man."—*English Mechanic.*
"Simply invaluable to the youngster entering upon this particular calling, and highly serviceable to the man who is practising it."—*Furniture Gazette.*

Decorators, Receipts for.
THE DECORATOR'S ASSISTANT: A Modern Guide to Decorative Artists and Amateurs, Painters, Writers, Gilders, &c. Containing upwards of 600 Receipts, Rules and Instructions; with a variety of Information for General Work connected with every Class of Interior and Exterior Decorations, &c. Fourth Edition, Revised. 152 pp., crown 8vo, 1s. in wrapper.

"Full of receipts of value to decorators, painters, gilders, &c. The book contains the gist of larger treatises on colour and technical processes. It would be difficult to meet with a work so full of varied information on the painter's art."—*Building News.*
"We recommend the work to all who, whether for pleasure or profit, require a guide to decoration."—*Plumber and Decorator.*

Moyr Smith on Interior Decoration.
ORNAMENTAL INTERIORS, ANCIENT AND MODERN. By J. MOYR SMITH. Super-royal 8vo, with 32 full-page Plates and numerous smaller Illustrations, handsomely bound in cloth, gilt top, price 18s.

"The book is well illustrated and handsomely got up, and contains some true criticism and a good many good examples of decorative treatment."—*The Builder.*
"This is the most elaborate and beautiful work on the artistic decoration of interiors that we have seen. . . . The scrolls, panels and other designs from the author's own pen are very beautiful and chaste; but he takes care that the designs of other men shall figure even more than his own."—*Liverpool Albion.*
"To all who take an interest in elaborate domestic ornament this handsome volume will be welcome."—*Graphic.*

British and Foreign Marbles.
MARBLE DECORATION and the *Terminology of British and Foreign Marbles.* A Handbook for Students. By GEORGE H. BLAGROVE, Author of "Shoring and its Application," &c. With 28 Illustrations. Crown 8vo, 3s. 6d. cloth.

"This most useful and much wanted handbook should be in the hands of every architect and builder."—*Building World.*
"It is an excellent manual for students, and interesting to artistic readers generally."—*Saturday Review.*
"A carefully and usefully written treatise; the work is essentially practical."—*Scotsman.*

Marble Working, etc.
MARBLE AND MARBLE WORKERS: A Handbook for Architects, Artists, Masons and Students. By ARTHUR LEE, Author of "A Visit to Carrara," "The Working of Marble," &c. Small crown 8vo, 2s. cloth.

"A really valuable addition to the technical literature of architects and masons."—*Building News.*

DELAMOTTE'S WORKS ON ILLUMINATION AND ALPHABETS.

A PRIMER OF THE ART OF ILLUMINATION, for the Use of Beginners: with a Rudimentary Treatise on the Art, Practical Directions for its exercise, and Examples taken from Illuminated MSS., printed in Gold and Colours. By F. DELAMOTTE. New and Cheaper Edition. Small 4to, 6s. ornamental boards.

"The examples of ancient MSS. recommended to the student, which, with much good sense, the author chooses from collections accessible to all, are selected with judgment and knowledge, as well as taste."—*Athenæum.*

ORNAMENTAL ALPHABETS, Ancient and Mediæval, from the Eighth Century, with Numerals; including Gothic, Church-Text, large and small, German, Italian, Arabesque, Initials for Illumination, Monograms, Crosses, &c. &c., for the use of Architectural and Engineering Draughtsmen, Missal Painters, Masons, Decorative Painters, Lithographers, Engravers, Carvers, &c. &c. Collected and Engraved by F. DELAMOTTE, and printed in Colours. New and Cheaper Edition. Royal 8vo, oblong, 2s. 6d. ornamental boards.

"For those who insert enamelled sentences round gilded chalices, who blazon shop legends over shop-doors, who letter church walls with pithy sentences from the Decalogue, this book will be useful."—*Athenæum.*

EXAMPLES OF MODERN ALPHABETS, Plain and Ornamental; including German, Old English, Saxon, Italic, Perspective, Greek, Hebrew, Court Hand, Engrossing, Tuscan, Riband, Gothic, Rustic, and Arabesque; with several Original Designs, and an Analysis of the Roman and Old English Alphabets, large and small, and Numerals, for the use of Draughtsmen, Surveyors, Masons, Decorative Painters, Lithographers, Engravers, Carvers, &c. Collected and Engraved by F. DELAMOTTE, and printed in Colours. New and Cheaper Edition. Royal 8vo, oblong, 2s. 6d. ornamental boards.

"There is comprised in it every possible shape into which the letters of the alphabet and numerals can be formed, and the talent which has been expended in the conception of the various plain and ornamental letters is wonderful."—*Standard.*

MEDIÆVAL ALPHABETS AND INITIALS FOR ILLUMINATORS. By F. G. DELAMOTTE. Containing 21 Plates and Illuminated Title, printed in Gold and Colours. With an Introduction by J. WILLIS BROOKS. Fourth and Cheaper Edition. Small 4to, 4s. ornamental boards.

"A volume in which the letters of the alphabet come forth glorified in gilding and all the colours of the prism interwoven and intertwined and intermingled."—*Sun.*

THE EMBROIDERER'S BOOK OF DESIGN. Containing Initials, Emblems, Cyphers, Monograms, Ornamental Borders, Ecclesiastical Devices, Mediæval and Modern Alphabets, and National Emblems. Collected by F. DELAMOTTE, and printed in Colours. Oblong royal 8vo, 1s. 6d. ornamental wrapper.

"The book will be of great assistance to ladies and young children who are endowed with the art of plying the needle in this most ornamental and useful pretty work."—*East Anglian Times.*

Wood Carving.
INSTRUCTIONS IN WOOD-CARVING, for Amateurs; with Hints on Design. By A LADY. With Ten Plates. New and Cheaper Edition. Crown 8vo, 2s. in emblematic wrapper.

"The handicraft of the wood-carver, so well as a book can impart it, may be learnt from 'A Lady's' publication."—*Athenæum.*
"The directions given are plain and easily understood."—*English Mechanic.*

Glass Painting.
GLASS STAINING AND THE ART OF PAINTING ON GLASS. From the German of Dr. GESSERT and EMANUEL OTTO FROMBERG. With an Appendix on THE ART OF ENAMELLING. 12mo, 2s. 6d. cloth limp.

Letter Painting.
THE ART OF LETTER PAINTING MADE EASY. By JAMES GREIG BADENOCH. With 12 full-page Engravings of Examples, 1s. 6d. cloth limp.

"The system is a simple ore, but quite original, and well worth the careful attention of letter painters. It can be easily mastered and remembered."—*Building News.*

CARPENTRY, TIMBER, etc.

Tredgold's Carpentry, Revised & Enlarged by Tarn.
THE ELEMENTARY PRINCIPLES OF CARPENTRY. A Treatise on the Pressure and Equilibrium of Timber Framing, the Resistance of Timber, and the Construction of Floors, Arches, Bridges, Roofs, Uniting Iron and Stone with Timber, &c. To which is added an Essay on the Nature and Properties of Timber, &c., with Descriptions of the kinds of Wood used in Building; also numerous Tables of the Scantlings of Timber for different purposes, the Specific Gravities of Materials, &c. By THOMAS TREDGOLD, C.E. With an Appendix of Specimens of Various Roofs of Iron and Stone, Illustrated. Seventh Edition, thoroughly revised and considerably enlarged by E. WYNDHAM TARN, M.A., Author of "The Science of Building," &c. With 61 Plates, Portrait of the Author, and several Woodcuts. In one large vol., 4to, price £1 5s. cloth.
"Ought to be in every architect's and every builder's library."—*Builder*.
"A work whose monumental excellence must commend it wherever skilful carpentry is concerned. The author's principles are rather confirmed than impaired by time. The additional plates are of great intrinsic value."—*Building News*.

Woodworking Machinery.
WOODWORKING MACHINERY : *Its Rise, Progress, and Construction.* With Hints on the Management of Saw Mills and the Economical Conversion of Timber. Illustrated with Examples of Recent Designs by leading English, French, and American Engineers. By M. POWIS BALE, A.M.Inst.C.E.,M.I.M.E. Large crown 8vo, 12s. 6d. cloth.
"Mr. Bale is evidently an expert on the subject and he has collected so much information that his book is all-sufficient for builders and others engaged in the conversion of timber."—*Architect*.
"The most comprehensive compendium of wood-working machinery we have seen. The author is a thorough master of his subject."—*Building News*.
"The appearance of this book at the present time will, we should think, give a considerable impetus to the onward march of the machinist engaged in the designing and manufacture of wood-working machines. It should be in the office of every wood-working factory."—*English Mechanic*.

Saw Mills.
SAW MILLS : *Their Arrangement and Management, and the Economical Conversion of Timber.* (A Companion Volume to " Woodworking Machinery.") By M. POWIS BALE. With numerous Illustrations. Crown 8vo, 10s. 6d. cloth.
"The *administration* of a large sawing establishment is discussed, and the subject examined from a financial standpoint. Hence the size, shape, order, and disposition of saw mills and the like are gone into in detail, and the course of the timber is traced from its reception to its delivery in its converted state. We could not desire a more complete or practical treatise."—*Builder*.
"We highly recommend Mr. Bale's work to the attention and perusal of all those who are engaged in the art of wood conversion, or who are about building or remodelling saw-mills on improved principles."—*Building News*.

Carpentering.
THE CARPENTER'S NEW GUIDE ; or, Book of Lines for Carpenters; comprising all the Elementary Principles essential for acquiring a knowledge of Carpentry. Founded on the late PETER NICHOLSON's Standard Work. A New Edition, Revised by ARTHUR ASHPITEL, F.S.A. Together with Practical Rules on Drawing, by GEORGE PYNE. With 74 Plates, 4to, £1 1s. cloth.

Handrailing and Stairbuilding.
A PRACTICAL TREATISE ON HANDRAILING : Showing New and Simple Methods for Finding the Pitch of the Plank, Drawing the Moulds, Bevelling, Jointing-up, and Squaring the Wreath. By GEORGE COLLINGS. Second Edition, Revised and Enlarged, to which is added A TREATISE ON STAIRBUILDING. With Plates and Diagrams. 12mo, 2s. 6d. cloth limp. [*Just published*.
"Will be found of practical utility in the execution of this difficult branch of joinery."—*Builder*.
"Almost every difficult phase of this somewhat intricate branch of joinery is elucidated by the aid of plates and explanatory letterpress."—*Furniture Gazette*.

Circular Work.
CIRCULAR WORK IN CARPENTRY AND JOINERY : A Practical Treatise on Circular Work of Single and Double Curvature. By GEORGE COLLINGS, Author of "A Practical Treatise on Handrailing." Illustrated with numerous Diagrams. Second Edition. 12mo, 2s. 6d. cloth limp.
"An excellent example of what a book of this kind should be. Cheap in price, clear in definition and practical in the examples selected."—*Builder*.

Timber Merchant's Companion.

THE TIMBER MERCHANT'S AND BUILDER'S COMPANION. Containing New and Copious Tables of the Reduced Weight and Measurement of Deals and Battens, of all sizes, from One to a Thousand Pieces, and the relative Price that each size bears per Lineal Foot to any given Price per Petersburg Standard Hundred; the Price per Cube Foot of Square Timber to any given Price per Load of 50 Feet; the proportionate Value of Deals and Battens by the Standard, to Square Timber by the Load of 50 Feet; the readiest mode of ascertaining the Price of Scantling per Lineal Foot of any size, to any given Figure per Cube Foot, &c. &c. By WILLIAM DOWSING. Fourth Edition, Revised and Corrected. Cr. 8vo, 3s. cl.

"Everything is as concise and clear as it can possibly be made. There can be no doubt that every timber merchant and builder ought to possess it."—*Hull Advertiser.*

"We are glad to see a fourth edition of these admirable tables, which for correctness and simplicity of arrangement leave nothing to be desired."—*Timber Trades Journal.*

"An exceedingly well-arranged, clear, and concise manual of tables for the use of all who buy or sell timber."—*Journal of Forestry.*

Practical Timber Merchant.

THE PRACTICAL TIMBER MERCHANT. Being a Guide for the use of Building Contractors, Surveyors, Builders, &c., comprising useful Tables for all purposes connected with the Timber Trade, Marks of Wood, Essay on the Strength of Timber, Remarks on the Growth of Timber, &c. By W. RICHARDSON. Fcap. 8vo, 3s. 6d. cloth.

"This handy manual contains much valuable information for the use of timber merchants, builders, foresters, and all others connected with the growth, sale, and manufacture of timber."—*Journal of Forestry.*

Timber Freight Book.

THE TIMBER MERCHANT'S, SAW MILLER'S, AND IMPORTER'S FREIGHT BOOK AND ASSISTANT. Comprising Rules, Tables, and Memoranda relating to the Timber Trade. By WILLIAM RICHARDSON, Timber Broker; together with a Chapter on "SPEEDS OF SAW MILL MACHINERY," by M. POWIS BALE, M.I.M.E., &c. 12mo, 3s. 6d. cl. boards.

"A very useful manual of rules, tables, and memoranda relating to the timber trade. We recommend it as a compendium of calculation to all timber measurers and merchants, and as supplying a real want in the trade."—*Building News.*

Packing-Case Makers, Tables for.

PACKING-CASE TABLES; showing the number of Superficial Feet in Boxes or Packing-Cases, from six inches square and upwards. By W. RICHARDSON, Timber Broker. Second Edition. Oblong 4to, 3s. 6d. cl.

"Invaluable labour-saving tables."—*Ironmonger.*
"Will save much labour and calculation."—*Grocer.*

Superficial Measurement.

THE TRADESMAN'S GUIDE TO SUPERFICIAL MEASUREMENT. Tables calculated from 1 to 200 inches in length, by 1 to 108 inches in breadth. For the use of Architects, Surveyors, Engineers, Timber Merchants, Builders, &c. By JAMES HAWKINGS. Third Edition. Fcap., 3s. 6d. cloth.

"A useful collection of tables to facilitate rapid calculation of surfaces. The exact area of any surface of which the limits have been ascertained can be instantly determined. The book will be found of the greatest utility to all engaged in building operations."—*Scotsman.*

"These tables will be found of great assistance to all who require to make calculations in superficial measurement."—*English Mechanic.*

Forestry.

THE ELEMENTS OF FORESTRY. Designed to afford Information concerning the Planting and Care of Forest Trees for Ornament or Profit, with Suggestions upon the Creation and Care of Woodlands. By F. B. HOUGH. Large crown 8vo, 10s. cloth.

Timber Importer's Guide.

THE TIMBER IMPORTER'S, TIMBER MERCHANT'S AND BUILDER'S STANDARD GUIDE. By RICHARD E. GRANDY. Comprising an Analysis of Deal Standards, Home and Foreign, with Comparative Values and Tabular Arrangements for fixing Nett Landed Cost on Baltic and North American Deals, including all intermediate Expenses, Freight, Insurance, &c. &c. Together with copious Information for the Retailer and Builder. Third Edition, Revised. 12mo, 2s. cloth limp.

"Everything it pretends to be: built up gradually, it leads one from a forest to a treenail, and throws in, as a makeweight, a host of material concerning bricks, columns, cisterns, &c."—*English Mechanic.*

MARINE ENGINEERING, NAVIGATION, etc.

Chain Cables.
CHAIN CABLES AND CHAINS. Comprising Sizes and Curves of Links, Studs, &c., Iron for Cables and Chains, Chain Cable and Chain Making, Forming and Welding Links, Strength of Cables and Chains, Certificates for Cables, Marking Cables, Prices of Chain Cables and Chains, Historical Notes, Acts of Parliament, Statutory Tests, Charges for Testing, List of Manufacturers of Cables, &c. &c. By THOMAS W. TRAILL, F.E.R.N., M. Inst. C.E., Engineer Surveyor in Chief, Board of Trade, Inspector of Chain Cable and Anchor Proving Establishments, and General Superintendent, Lloyd's Committee on Proving Establishments. With numerous Tables, Illustrations and Lithographic Drawings. Folio, £2 2s. cloth, bevelled boards.

"It contains a vast amount of valuable information. Nothing seems to be wanting to make it a complete and standard work of reference on the subject."—*Nautical Magazine.*

Marine Engineering.
MARINE ENGINES AND STEAM VESSELS *(A Treatise on).* By ROBERT MURRAY, C.E. Eighth Edition, thoroughly Revised, with considerable Additions by the Author and by GEORGE CARLISLE, C.E., Senior Surveyor to the Board of Trade at Liverpool. 12mo, 5s. cloth boards.

"Well adapted to give the young steamship engineer or marine engine and boiler maker a general introduction into his practical work."—*Mechanical World.*

"We feel sure that this thoroughly revised edition will continue to be as popular in the future as it has been in the past, as, for its size, it contains more useful information than any similar treatise."—*Industries.*

"As a compendious and useful guide to engineers of our mercantile and royal naval services we should say it cannot be surpassed."—*Building News.*

"The information given is both sound and sensible, and well qualified to direct young seagoing hands on the straight road to the extra chief's certificate. Most useful to surveyors, inspectors, draughtsmen, and all young engineers who take an interest in their profession.—*Glasgow Herald.*

"An indispensable manual for the student of marine engineering."—*Liverpool Mercury.*

Pocket-Book for Naval Architects and Shipbuilders.
THE NAVAL ARCHITECT'S AND SHIPBUILDER'S POCKET-BOOK of Formulæ, Rules, and Tables, and MARINE ENGINEER'S AND SURVEYOR'S Handy Book of Reference. By CLEMENT MACKROW, Member of the Institution of Naval Architects, Naval Draughtsman. Fourth Edition, Revised. With numerous Diagrams, &c. Fcap., 12s. 6d. strongly bound in leather.

"Should be used by all who are engaged in the construction or design of vessels. . . . Will be found to contain the most useful tables and formulæ required by shipbuilders, carefully collected from the best authorities, and put together in a popular and simple form."—*Engineer.*

"The professional shipbuilder has now, in a convenient and accessible form, reliable data for solving many of the numerous problems that present themselves in the course of his work."—*Iron.*

"There is scarcely a subject on which a naval architect or shipbuilder can require to refresh his memory which will not be found within the covers of Mr. Mackrow's book."—*English Mechanic.*

Pocket-Book for Marine Engineers.
A POCKET-BOOK OF USEFUL TABLES AND FORMULÆ FOR MARINE ENGINEERS. By FRANK PROCTOR, A.I.N.A. Third Edition. Royal 32mo, leather, gilt edges, with strap, 4s.

"We recommend it to our readers as going far to supply a long-felt want."—*Naval Science.*
"A most useful companion to all marine engineers."—*United Service Gazette.*

Introduction to Marine Engineering.
ELEMENTARY ENGINEERING: A Manual for Young Marine Engineers and Apprentices. In the Form of Questions and Answers on Metals, Alloys, Strength of Materials, Construction and Management of Marine Engines and Boilers, Geometry, &c. &c. With an Appendix of Useful Tables. By JOHN SHERREN BREWER, Government Marine Surveyor, Hongkong. Small crown 8vo, 2s. cloth.

"Contains much valuable information for the class for whom it is intended, especially in the chapters on the management of boilers and engines."—*Nautical Magazine.*
"A useful introduction to the more elaborate text books."—*Scotsman.*
"To a student who has the requisite desire and resolve to attain a thorough knowledge, Mr. Brewer offers decidedly useful help."—*Athenæum.*

Navigation.
PRACTICAL NAVIGATION. Consisting of THE SAILOR'S SEA-BOOK, by JAMES GREENWOOD and W. H. ROSSER; together with the requisite Mathematical and Nautical Tables for the Working of the Problems, by HENRY LAW, C.E., and Professor J. R. YOUNG. Illustrated, 12mo, 7s, strongly half-bound.

MINING AND METALLURGY.

Metalliferous Mining in the United Kingdom.
BRITISH MINING: A Treatise on the History, Discovery, Practical Development, and Future Prospects of Metalliferous Mines in the United Kingdom. By ROBERT HUNT, F.R.S., Keeper of Mining Records; Editor of "Ure's Dictionary of Arts, Manufactures, and Mines," &c. Upwards of 950 pp., with 230 Illustrations. Second Edition, Revised. Super-royal 8vo, £2 2s. cloth.
"One of the most valuable works of reference of modern times. Mr. Hunt, as keeper of mining records of the United Kingdom, has had opportunities for such a task not enjoyed by anyone else, and has evidently made the most of them. . . . The language and style adopted are good, and the treatment of the various subjects laborious, conscientious, and scientific."—*Engineering.*
"The book is, in fact, a treasure-house of statistical information on mining subjects, and we know of no other work embodying so great a mass of matter of this kind. Were this the only merit of Mr. Hunt's volume, it would be sufficient to render it indispensable in this library of everyone interested in the development of the mining and metallurgical industries of this country."—*Athenæum.*
"A mass of information not elsewhere available, and of the greatest value to those who may be interested in our great mineral industries."—*Engineer.*
"A sound, business-like collection of interesting facts. . . . The amount of information Mr. Hunt has brought together is enormous. . . . The volume appears likely to convey more instruction upon the subject than any work hitherto published."—*Mining Journal.*

Colliery Management.
THE COLLIERY MANAGER'S HANDBOOK: A Comprehensive Treatise on the Laying-out and Working of Collieries, Designed as a Book of Reference for Colliery Managers, and for the Use of Coal-Mining Students preparing for First-class Certificates. By CALEB PAMELY, Mining Engineer and Surveyor; Member of the North of England Institute of Mining and Mechanical Engineers; and Member of the South Wales Institute of Mining Engineers. With nearly 500 Plans, Diagrams, and other Illustrations. Medium 8vo, about 600 pages. Price £1 5s. strongly bound.
[*Just ready.*

Coal and Iron.
THE COAL AND IRON INDUSTRIES OF THE UNITED KINGDOM. Comprising a Description of the Coal Fields, and of the Principal Seams of Coal, with Returns of their Produce and its Distribution, and Analyses of Special Varieties. Also an Account of the occurrence of Iron Ores in Veins or Seams; Analyses of each Variety; and a History of the Rise and Progress of Pig Iron Manufacture. By RICHARD MEADE, Assistant Keeper of Mining Records. With Maps. 8vo, £1 8s. cloth.
"The book is one which must find a place on the shelves of all interested in coal and iron production, and in the iron, steel, and other metallurgical industries."—*Engineer.*
"Of this book we may unreservedly say that it is the best of its class which we have ever met. . . . A book of reference which no one engaged in the iron or coal trades should omit from his library."—*Iron and Coal Trades Review.*

Prospecting for Gold and other Metals.
THE PROSPECTOR'S HANDBOOK: A Guide for the Prospector and Traveller in Search of Metal-Bearing or other Valuable Minerals. By J. W. ANDERSON, M.A. (Camb.), F.R.G.S., Author of "Fiji and New Caledonia." Fifth Edition, thoroughly Revised and Enlarged. Small crown 8vo, 3s. 6d. cloth.
"Will supply a much felt want, especially among Colonists, in whose way are so often thrown many mineralogical specimens the value of which it is difficult to determine."—*Engineer.*
"How to find commercial minerals, and how to identify them when they are found, are the leading points to which attention is directed. The author has managed to pack as much practical detail into his pages as would supply material for a book three times its size."—*Mining Journal.*

Mining Notes and Formulæ.
NOTES AND FORMULÆ FOR MINING STUDENTS. By JOHN HERMAN MERIVALE, M.A., Certificated Colliery Manager, Professor of Mining in the Durham College of Science, Newcastle-upon-Tyne. Third Edition, Revised and Enlarged. Small crown 8vo, 2s. 6d. cloth. [*Just published.*
"Invaluable to anyone who is working up for an examination on mining subjects."—*Coal and Iron Trades Review.*
"The author has done his work in an exceedingly creditable manner, and has produced a book that will be of service to students, and those who are practically engaged in mining operations."—*Engineer.*
"A vast amount of technical matter of the utmost value to mining engineers, and of considerable interest to students."—*Schoolmaster.*

Explosives.

A HANDBOOK ON MODERN EXPLOSIVES. Being a Practical Treatise on the Manufacture and Application of Dynamite, Gun-Cotton, Nitro-Glycerine and other Explosive Compounds. Including the Manufacture of Collodion-Cotton. By M. EISSLER, Mining Engineer and Metallurgical Chemist, Author of "The Metallurgy of Gold," "The Metallurgy of Silver," &c. With about 100 Illustrations. Crown 8vo, 10s. 6d. cloth. [*Just published*.

"Useful not only to the miner, but also to officers of both services to whom blasting and the use of explosives generally may at any time become a necessary auxiliary."—*Nature*.

"A veritable mine of information on the subject of explosives employed for military, mining and blasting purposes."—*Army and Navy Gazette*.

"The book is clearly written. Taken as a whole, we consider it an excellent little book and one that should be found of great service to miners and others who are engaged in work requiring the use of explosives."—*Athenæum*.

Gold, Metallurgy of.

THE METALLURGY OF GOLD: A Practical Treatise on the Metallurgical Treatment of Gold-bearing Ores. Including the Processes of Concentration and Chlorination, and the Assaying, Melting and Refining of Gold. By M. EISSLER, Mining Engineer and Metallurgical Chemist, formerly Assistant Assayer of the U. S. Mint, San Francisco. Third Edition, Revised and greatly Enlarged. With 187 Illustrations. Crown 8vo, 12s. 6d. cloth. [*Just published*.

"This book thoroughly deserves its title of a 'Practical Treatise.' The whole process of gold milling, from the breaking of the quartz to the assay of the bullion, is described in clear and orderly narrative and with much, but not too much, fulness of detail."—*Saturday Review*.

"The work is a storehouse of information and valuable data, and we strongly recommend it to all professional men engaged in the gold-mining industry."—*Mining Journal*.

Silver, Metallurgy of.

THE METALLURGY OF SILVER: A Practical Treatise on the Amalgamation, Roasting and Lixiviation of Silver Ores. Including the Assaying, Melting and Refining of Silver Bullion. By M. EISSLER, Author of "The Metallurgy of Gold." With 124 Illustrations. Crown 8vo, 10s. 6d. cloth.

"A practical treatise, and a technical work which we are convinced will supply a long-felt want amongst practical men, and at the same time be of value to students and others indirectly connected with the industries."—*Mining Journal*.

"From first to last the book is thoroughly sound and reliable."—*Colliery Guardian*.

"For chemists, practical miners, assayers and investors alike, we do not know of any work on the subject so handy and yet so comprehensive."—*Glasgow Herald*.

Silver-Lead, Metallurgy of.

THE METALLURGY OF ARGENTIFEROUS LEAD ORES: A Practical Treatise on the Smelting of Silver-Lead Ores and the Refining of Lead Bullion. Illustrated with Plans and Sections of Smelting Furnaces and Plant in Europe and America. By M. EISSLER, Author of "The Metallurgy of Gold," "The Metallurgy of Silver," &c. Cr. 8vo. [*In the press*.

Metalliferous Minerals and Mining.

TREATISE ON METALLIFEROUS MINERALS AND MINING. By D. C. DAVIES, F.G.S., Mining Engineer, &c., Author of "A Treatise on Slate and Slate Quarrying." Illustrated with numerous Wood Engravings. Fourth Edition, carefully Revised. Crown 8vo, 12s. 6d. cloth.

"Neither the practical miner nor the general reader interested in mines can have a better book for his companion and his guide."—*Mining Journal*.

"We are doing our readers a service in calling their attention to this valuable work."—*Mining World*.

"A book that will not only be useful to the geologist, the practical miner, and the metallurgist, but also very interesting to the general public."—*Iron*.

"As a history of the present state of mining throughout the world this book has a real value, and it supplies an actual want."—*Athenæum*.

Earthy Minerals and Mining.

A TREATISE ON EARTHY & OTHER MINERALS AND MINING. By D. C. DAVIES, F.G.S. Uniform with, and forming a Companion Volume to, the same Author's "Metalliferous Minerals and Mining." With 76 Wood Engravings. Second Edition. Crown 8vo, 12s. 6d. cloth.

"We do not remember to have met with any English work on mining matters that contains the same amount of information packed in equally convenient form."—*Academy*.

"We should be inclined to rank it as among the very best of the handy technical and trades manuals which have recently appeared."—*British Quarterly Review*.

Mineral Surveying and Valuing.

THE MINERAL SURVEYOR AND VALUER'S COMPLETE GUIDE, comprising a Treatise on Improved Mining Surveying and the Valuation of Mining Properties, with New Traverse Tables. By WM. LINTERN, Mining and Civil Engineer. Third Edition, with an Appendix on "Magnetic and Angular Surveying," with Records of the Peculiarities of Needle Disturbances. With Four Plates of Diagrams, Plans, &c. 12mo, 4s. cloth.
[*Just published.*

"Mr. Lintern's book forms a valuable and thoroughly trustworthy guide."—*Iron and Coal Trades Review.*
"This new edition must be of the highest value to colliery surveyors, proprietors and managers."—*Colliery Guardian.*

Asbestos and its Uses.

ASBESTOS: Its Properties, Occurrence and Uses. With some Account of the Mines of Italy and Canada. By ROBERT H. JONES. With Eight Collotype Plates and other Illustrations. Crown 8vo, 12s. 6d. cloth.
[*Just published.*

"An interesting and invaluable work."—*Colliery Guardian.*
"We counsel our readers to get this exceedingly interesting work for themselves; they will find in it much that is suggestive, and a great deal that is of immediate and practical usefulness."—*Builder.*
"A valuable addition to the architect's and engineer's library."—*Building News.*

Underground Pumping Machinery.

MINE DRAINAGE. Being a Complete and Practical Treatise on Direct-Acting Underground Steam Pumping Machinery, with a Description of a large number of the best known Engines, their General Utility and the Special Sphere of their Action, the Mode of their Application, and their merits compared with other forms of Pumping Machinery. By STEPHEN MICHELL. 8vo, 15s. cloth.

"Will be highly esteemed by colliery owners and lessees, mining engineers, and students generally who require to be acquainted with the best means of securing the drainage of mines. It is a most valuable work, and stands almost alone in the literature of steam pumping machinery."—*Colliery Guardian.*
"Much valuable information is given, so that the book is thoroughly worthy of an extensive circulation amongst practical men and purchasers of machinery."—*Mining Journal.*

Mining Tools.

A MANUAL OF MINING TOOLS. For the Use of Mine Managers, Agents, Students, &c. By WILLIAM MORGANS, Lecturer on Practical Mining at the Bristol School of Mines. 12mo, 2s. 6d. cloth limp.

ATLAS OF ENGRAVINGS to Illustrate the above, containing 235 Illustrations of Mining Tools, drawn to scale. 4to, 4s. 6d. cloth.
"Students in the science of mining, and overmen, captains, managers, and viewers may gain practical knowledge and useful hints by the study of Mr. Morgans' manual."—*Colliery Guardian.*
"A valuable work, which will tend materially to improve our mining literature."—*Mining Journal.*

Coal Mining.

COAL AND COAL MINING: A Rudimentary Treatise on. By the late Sir WARINGTON W. SMYTH, M.A., F.R.S., &c., Chief Inspector of the Mines of the Crown. Seventh Edition, Revised and Enlarged. With numerous Illustrations. 12mo, 4s. cloth boards. [*Just published.*
"As an outline is given of every known coal-field in this and other countries, as well as of the principal methods of working, the book will doubtless interest a very large number of readers."—*Mining Journal.*

Subterraneous Surveying.

SUBTERRANEOUS SURVEYING, Elementary and Practical Treatise on, with and without the Magnetic Needle. By THOMAS FENWICK, Surveyor of Mines, and THOMAS BAKER, C.E. Illust. 12mo, 3s. cloth boards.

Granite Quarrying.

GRANITES AND OUR GRANITE INDUSTRIES. By GEORGE F. HARRIS, F.G.S., Membre de la Société Belge de Géologie, Lecturer on Economic Geology at the Birkbeck Institution, &c. With Illustrations. Crown 8vo, 2s. 6d. cloth.
"A clearly and well-written manual for persons engaged or interested in the granite industry."—*Scotsman.*
"An interesting work, which will be deservedly esteemed."—*Colliery Guardian.*
"An exceedingly interesting and valuable monograph on a subject which has hitherto received unaccountably little attention in the shape of systematic literary treatment."—*Scottish Leader.*

ELECTRICITY, ELECTRICAL ENGINEERING, etc.

ELECTRICITY, ELECTRICAL ENGINEERING, etc.

Electrical Engineering.

THE ELECTRICAL ENGINEER'S POCKET-BOOK OF MODERN RULES, FORMULÆ, TABLES AND DATA. By H. R. KEMPE, M.Inst.E.E., A.M.Inst C.E., Technical Officer Postal Telegraphs, Author of "A Handbook of Electrical Testing," &c. With numerous Illustrations, royal 32mo, oblong, 5s. leather. [*Just published.*

"There is very little in the shape of formulæ or data which the electrician is likely to want in a hurry which cannot be found in its pages."—*Practical Engineer.*

"A very useful book of reference for daily use in practical electrical engineering and its various applications to the industries of the present day."—*Iron.*

"It is the best book of its kind."—*Electrical Engineer.*

"We l arranged and compact. The Electrical Engineer's Pocket-Book is a good one."—*Electrician.*

"Strongly recommended to those engaged in the various electrical industries."—*Electrical Review.*

Electric Lighting.

ELECTRIC LIGHT FITTING: A Handbook for Working Electrical Engineers, embodying Practical Notes on Installation Management. By JOHN W. URQUHART, Electrician, Author of "Electric Light," &c. With numerous Illustrations, crown 8vo, 5s. cloth. [*Just published.*

"This volume deals with what may be termed the mechanics of electric lighting, and is addressed to men who are already engaged in the work or are training for it. The work traverses a great deal of ground, and may be read as a sequel to the same author's useful work on 'Electric Light.'"—*Electrician.*

"This is an attempt to state in the simplest language the precautions which should be adopted in installing the electric light, and to give information,for the guidance of those who have to run the plant when installed. The book is well worth the perusal of the workmen for whom it is written."—*Electrical Review.*

"Eminently practical and useful. . . . Ought to be in the hands of everyone in charge of an electric light plant."—*Electrical Engineer.*

"Altogether Mr. Urquhart has succeeded in producing a really capital book, which we have no hesitation in recommending to the notice of working electricians and electrical engineers.—*Mechanical World.*

Electric Light.

ELECTRIC LIGHT: Its Production and Use. Embodying Plain Directions for the Treatment of Dynamo-Electric Machines, Batteries, Accumulators, and Electric Lamps. By J. W. URQUHART, C.E., Author of "Electric Light Fitting," "Electroplating," &c. Fourth Edition, carefully Revised, with Large Additions and 145 Illustrations. Crown 8vo, 7s. 6d. cloth. [*Just published.*

"The book is by far the best that we have yet met with on the subject."—*Athenæum.*

"It is the only work at present available which gives, in language intelligible for the most part to the ordinary reader, a general but concise history of the means which have been adopted up to the present time in producing the electric light."—*Metropolitan.*

"The book contains a general account of the means adopted in producing the electric light, not only as obtained from voltaic or galvanic batteries, but treats at length of the dynamo-electric machine in several of its forms."—*Colliery Guardian.*

Construction of Dynamos.

DYNAMO CONSTRUCTION: A Practical Handbook for the Use of Engineer Constructors and Electricians in Charge. With Examples of leading English, American and Continental Dynamos and Motors. By J. W. URQUHART, Author of "Electric Light," "Electric Light Fitting," &c. Crown 8vo. [*In the press.*

Text Book of Electricity.

THE STUDENT'S TEXT-BOOK OF ELECTRICITY. By HENRY M. NOAD, Ph.D., F.R.S., F.C.S. New Edition, carefully Revised. With an Introduction and Additional Chapters, by W. H. PREECE, M.I.C.E., Vice-President of the Society of Telegraph Engineers, &c. With 470 Illustrations. Crown 8vo, 12s. 6d. cloth.

"The original plan of this book has been carefully adhered to so as to make it a reflex of the existing state of electrical science, adapted for students. . . . Discovery seems to have progressed with marvellous strides; nevertheless it has now apparently ceased, and practical applications have commenced their career; and it is to give a faithful account of these that this fresh edition of Dr. Noad's valuable text-book is launched forth."—*Extract from Introduction by W. H. Preece, Esq.*

"We can recommend Dr. Noad's book for clear style, great range of subject, a good index and a plethora of woodcuts. Such collections as the present are indispensable."—*Athenæum.*

"An admirable text book for every student — beginner or advanced — of electricity."—*Engineering.*

Electric Lighting.

THE ELEMENTARY PRINCIPLES OF ELECTRIC LIGHT-ING. By ALAN A. CAMPBELL SWINTON, Associate I.E.E. Second Edition, Enlarged and Revised. With 16 Illustrations. Crown 8vo, 1s. 6d. cloth.

"Anyone who desires a short and thoroughly clear exposition of the elementary principles of electric-lighting cannot do better than read this little work."—*Bradford Observer.*

Electricity.

A MANUAL OF ELECTRICITY: *Including Galvanism, Magnetism, Dia-Magnetism, Electro-Dynamics, Magno-Electricity, and the Electric Telegraph.* By HENRY M. NOAD, Ph.D., F.R.S., F.C.S. Fourth Edition. With 500 Woodcuts. 8vo, £1 4s. cloth.

"It is worthy of a place in the library of every public institution."—*Mining Journal.*

Dynamo Construction.

HOW TO MAKE A DYNAMO: *A Practical Treatise for Amateurs.* Containing numerous Illustrations and Detailed Instructions for Constructing a Small Dynamo, to Produce the Electric Light. By ALFRED CROFTS. Third Edition, Revised and Enlarged. Crown 8vo, 2s. cloth. [*Just published.*

"The instructions given in this unpretentious little book are sufficiently clear and explicit to enable any amateur mechanic possessed of average skill and the usual tools to be found in an amateur's workshop, to build a practical dynamo machine."—*Electrician.*

NATURAL SCIENCE, etc.

Pneumatics and Acoustics.

PNEUMATICS: *including Acoustics and the Phenomena of Wind Currents*, for the Use of Beginners. By CHARLES TOMLINSON, F.R.S., F.C.S., &c. Fourth Edition, Enlarged. With numerous Illustrations. 12mo, 1s. 6d. cloth.

"Beginners in the study of this important application of science could not have a better manual."—*Scotsman.*

"A valuable and suitable text-book for students of Acoustics and the Phenomena of Wind Currents."—*Schoolmaster.*

Conchology.

A MANUAL OF THE MOLLUSCA: *Being a Treatise on Recent and Fossil Shells.* By S. P. WOODWARD, A.L.S., F.G.S., late Assistant Palæontologist in the British Museum. With an Appendix on *Recent and Fossil Conchological Discoveries*, by RALPH TATE, A.L.S., F.G.S. Illustrated by A. N. WATERHOUSE and JOSEPH WILSON LOWRY. With 23 Plates and upwards of 300 Woodcuts. Reprint of Fourth Ed., 1880. Cr. 8vo, 7s. 6d. cl.

"A most valuable storehouse of conchological and geological information."—*Science Gossip.*

Geology.

RUDIMENTARY TREATISE ON GEOLOGY, PHYSICAL AND HISTORICAL. Consisting of "Physical Geology," which sets forth the leading Principles of the Science; and "Historical Geology," which treats of the Mineral and Organic Conditions of the Earth at each successive epoch, especial reference being made to the British Series of Rocks. By RALPH TATE, A.L.S., F.G.S., &c. &c. With 250 Illustrations. 12mo, 5s. cloth boards.

"The fulness of the matter has elevated the book into a manual. Its information is exhaustive and well arranged."—*School Board Chronicle.*

Geology and Genesis.

THE TWIN RECORDS OF CREATION: *or, Geology and Genesis: their Perfect Harmony and Wonderful Concord.* By GEORGE W. VICTOR LE VAUX. Numerous Illustrations. Fcap. 8vo, 5s. cloth.

"A valuable contribution to the evidences of Revelation, and disposes very conclusively of the arguments of those who would set God's Works against God's Word. No real difficulty is shirked, and no sophistry is left unexposed."—*The Rock.*

"The remarkable peculiarity of this author is that he combines an unbounded admiration of science with an unbounded admiration of the Written record. The two impulses are balanced to a nicety; and the consequence is that difficulties, which to minds less evenly poised would be serious, find immediate solutions of the happiest kinds."—*London Review.*

Astronomy.

ASTRONOMY. By the late Rev. ROBERT MAIN, M.A., F.R.S., formerly Radcliffe Observer at Oxford. Third Edition, Revised and Corrected to the present time, by WILLIAM THYNNE LYNN, B.A., F.R.A.S., formerly of the Royal Observatory, Greenwich. 12mo, 2s. cloth limp.

"A sound and simple treatise, very carefully edited, and a capital book for beginners."—*Knowledge.* [*tional Times.*

"Accurately brought down to the requirements of the present time by Mr. Lynn."—*Educa-*

DR. LARDNER'S COURSE OF NATURAL PHILOSOPHY.

THE HANDBOOK OF MECHANICS. Enlarged and almost re-written by BENJAMIN LOEWY, F.R.A.S. With 378 Illustrations. Post 8vo, 6s. cloth.

"The perspicuity of the original has been retained, and chapters which had become obsolete have been replaced by others of more modern character. The explanations throughout are studiously popular, and care has been taken to show the application of the various branches of physics to the industrial arts, and to the practical business of life."—*Mining Journal.*

"Mr. Loewy has carefully revised the book, and brought it up to modern requirements."—*Nature.*

"Natural philosophy has had few exponents more able or better skilled in the art of popularising the subject than Dr. Lardner; and Mr. Loewy is doing good service in fitting this treatise, and the others of the series, for use at the present time."—*Scotsman.*

THE HANDBOOK OF HYDROSTATICS AND PNEUMATICS. New Edition, Revised and Enlarged, by BENJAMIN LOEWY, F.R.A.S. With 236 Illustrations. Post 8vo, 5s. cloth.

"For those 'who desire to attain an accurate knowledge of physical science without the profound methods of mathematical investigation,' this work is not merely intended, but well adapted."—*Chemical News.*

"The volume before us has been carefully edited, augmented to nearly twice the bulk of the former edition, and all the most recent matter has been added. . . . It is a valuable text-book."—*Nature.*

"Candidates for pass examinations will find it, we think, specially suited to their requirements."—*English Mechanic.*

THE HANDBOOK OF HEAT. Edited and almost entirely re-written by BENJAMIN LOEWY, F.R.A.S., &c. 117 Illustrations. Post 8vo, 6s. cloth.

"The style is always clear and precise, and conveys instruction without leaving any cloudiness or lurking doubts behind."—*Engineering.*

"A most exhaustive book on the subject on which it treats, and is so arranged that it can be understood by all who desire to attain an accurate knowledge of physical science. . . . . Mr. Loewy has included all the latest discoveries in the varied laws and effects of heat."—*Standard.*

"A complete and handy text-book for the use of students and general readers."—*English Mechanic.*

THE HANDBOOK OF OPTICS. By DIONYSIUS LARDNER, D.C.L., formerly Professor of Natural Philosophy and Astronomy in University College, London. New Edition. Edited by T. OLVER HARDING, B.A. Lond., of University College, London. With 298 Illustrations. Small 8vo, 448 pages, 5s. cloth.

"Written by one of the ablest English scientific writers, beautifully and elaborately illustrated."—*Mechanic's Magazine.*

THE HANDBOOK OF ELECTRICITY, MAGNETISM, AND ACOUSTICS. By Dr. LARDNER. Ninth Thousand. Edit. by GEORGE CAREY FOSTER, B.A., F.C.S. With 400 Illustrations. Small 8vo, 5s. cloth.

"The book could not have been entrusted to anyone better calculated to preserve the terse and lucid style of Lardner, while correcting his errors and bringing up his work to the present state of scientific knowledge."—*Popular Science Review.*

THE HANDBOOK OF ASTRONOMY. Forming a Companion to the "Handbook of Natural Philosophy." By DIONYSIUS LARDNER, D.C.L., formerly Professor of Natural Philosophy and Astronomy in University College, London. Fourth Edition. Revised and Edited by EDWIN DUNKIN, F.R.A.S., Royal Observatory, Greenwich. With 38 Plates and upwards of 100 Woodcuts. In One Vol., small 8vo, 550 pages, 9s. 6d. cloth.

"Probably no other book contains the same amount of information in so compendious and well-arranged a form—certainly none at the price at which this is offered to the public."—*Athenæum.*

"We can do no other than pronounce this work a most valuable manual of astronomy, and we strongly recommend it to all who wish to acquire a general—but at the same time correct—acquaintance with this sublime science."—*Quarterly Journal of Science.*

"One of the most deservedly popular books on the subject . . . We would recommend not only the student of the elementary principles of the science, but he who aims at mastering the higher and mathematical branches of astronomy, not to be without this work beside him."—*Practical Magazine.*

Dr. Lardner's Electric Telegraph.

THE ELECTRIC TELEGRAPH. By Dr. LARDNER. Revised and Re-written by E. B. BRIGHT, F.R.A.S. 140 Illustrations. Small 8vo, 2s. 6d. cloth.

"One of the most readable books extant on the Electric Telegraph."—*English Mechanic.*

DR. LARDNER'S MUSEUM OF SCIENCE AND ART.

THE MUSEUM OF SCIENCE AND ART. Edited by DIONYSIUS LARDNER, D.C.L., formerly Professor of Natural Philosophy and Astronomy in University College, London. With upwards of 1,200 Engravings on Wood. In 6 Double Volumes, £1 1s., in a new and elegant cloth binding; or handsomely bound in half-morocco, 31s. 6d.

*** OPINIONS OF THE PRESS.

"This series, besides affording popular but sound instruction on scientific subjects, with which the humblest man in the country ought to be acquainted, also undertakes that teaching of 'Common Things' which every well-wisher of his kind is anxious to promote. Many thousand copies of this serviceable publication have been printed, in the belief and hope that the desire for instruction and improvement widely prevails; and we have no fear that such enlightened faith will meet with disappointment."—*Times.*

"A cheap and interesting publication, alike informing and attractive. The papers combine subjects of importance and great scientific knowledge, considerable inductive powers, and a popular style of treatment."—*Spectator.*

"The 'Museum of Science and Art' is the most valuable contribution that has ever been made to the Scientific Instruction of every class of society."—Sir DAVID BREWSTER, in the *North British Review.*

"Whether we consider the liberality and beauty of the illustrations, the charm of the writing, or the durable interest of the matter, we must express our belief that there is hardly to be found among the new books one that would be welcomed by people of so many ages and classes as a valuable present."—*Examiner.*

*** *Separate books formed from the above, suitable for Workmen's Libraries, Science Classes, etc.*

Common Things Explained. Containing Air, Earth, Fire, Water, Time, Man, the Eye, Locomotion, Colour, Clocks and Watches, &c. 233 Illustrations, cloth gilt, 5s.

The Microscope. Containing Optical Images, Magnifying Glasses, Origin and Description of the Microscope, Microscopic Objects, the Solar Microscope, Microscopic Drawing and Engraving, &c. 147 Illustrations, cloth gilt, 2s.

Popular Geology. Containing Earthquakes and Volcanoes, the Crust of the Earth, &c. 201 Illustrations, cloth gilt, 2s. 6d.

Popular Physics. Containing Magnitude and Minuteness, the Atmosphere, Meteoric Stones, Popular Fallacies, Weather Prognostics, the Thermometer, the Barometer, Sound, &c. 85 Illustrations, cloth gilt, 2s. 6d.

Steam and its Uses. Including the Steam Engine, the Locomotive, and Steam Navigation. 89 Illustrations, cloth gilt, 2s.

Popular Astronomy. Containing How to observe the Heavens—The Earth, Sun, Moon, Planets, Light, Comets, Eclipses, Astronomical Influences, &c. 182 Illustrations, 4s. 6d.

The Bee and White Ants: Their Manners and Habits. With Illustrations of Animal Instinct and Intelligence. 135 Illustrations, cloth gilt, 2s.

The Electric Telegraph Popularized. To render intelligible to all who can Read, irrespective of any previous Scientific Acquirements, the various forms of Telegraphy in Actual Operation. 100 Illustrations, cloth gilt, 1s. 6d.

Dr. Lardner's School Handbooks.

NATURAL PHILOSOPHY FOR SCHOOLS. By Dr. LARDNER. 328 Illustrations. Sixth Edition. One Vol., 3s. 6d. cloth.

"A very convenient class-book for junior students in private schools. It is intended to convey, in clear and precise terms, general notions of all the principal divisions of Physical Science."—*British Quarterly Review.*

ANIMAL PHYSIOLOGY FOR SCHOOLS. By Dr. LARDNER. With 190 Illustrations. Second Edition. One Vol., 3s. 6d. cloth.

"Clearly written, well arranged, and excellently illustrated."—*Gardener's Chronicle.*

COUNTING-HOUSE WORK, TABLES, etc.

Accounts for Manufacturers.
FACTORY ACCOUNTS: Their Principles and Practice. A Handbook for Accountants and Manufacturers, with Appendices on the Nomenclature of Machine Details; the Income Tax Acts; the Rating of Factories; Fire and Boiler Insurance; the Factory and Workshop Acts, &c., including also a Glossary of Terms and a large number of Specimen Rulings. By EMILE GARCKE and J. M. FELLS. Third Edition. Demy 8vo, 250 pages, price 6s. strongly bound.

"A very interesting description of the requirements of Factory Accounts. . . . the principle of assimilating the Factory Accounts to the general commercial books is one which we thoroughly agree with."—*Accountants' Journal.*

"Characterised by extreme thoroughness. There are few owners of Factories who would not derive great benefit from the perusal of this most admirable work."—*Local Government Chronicle.*

Foreign Commercial Correspondence.
THE FOREIGN COMMERCIAL CORRESPONDENT: Being Aids to Commercial Correspondence in Five Languages—English, French, German, Italian and Spanish. By CONRAD E. BAKER. Second Edition, Revised. Crown 8vo, 3s. 6d. cloth.

"Whoever wishes to correspond in all the languages mentioned by Mr. Baker cannot do better than study this work, the materials of which are excellent and conveniently arranged. They consist not of entire specimen letters, but what are far more useful—short passages, sentences, or phrases expressing the same general idea in various forms."—*Athenæum.*

"A careful examination has convinced us that it is unusually complete, well arranged and reliable. The book is a thoroughly good one."—*Schoolmaster.*

Intuitive Calculations.
THE COMPENDIOUS CALCULATOR; or, Easy and Concise Methods of Performing the various Arithmetical Operations required in Commercial and Business Transactions, together with Useful Tables. By DANIEL O'GORMAN. Corrected and Extended by J. R. YOUNG, formerly Professor of Mathematics at Belfast College. Twenty-seventh Edition, carefully Revised by C. NORRIS. Fcap. 8vo, 2s. 6d. cloth limp; or, 3s. 6d. strongly half-bound in leather.

"It would be difficult to exaggerate the usefulness of a book like this to everyone engaged in commerce or manufacturing industry. It is crammed full of rules and formulæ for shortening and employing calculations."—*Knowledge.*

"Supplies special and rapid methods for all kinds of calculations. Of great utility to persons engaged in any kind of commercial transactions."—*Scotsman.*

Modern Metrical Units and Systems.
MODERN METROLOGY: A Manual of the Metrical Units and Systems of the Present Century. With an Appendix containing a proposed English System. By LOWIS D'A. JACKSON, A.M.Inst.C.E., Author of "Aid to Survey Practice," &c. Large crown 8vo, 12s. 6d. cloth.

"The author has brought together much valuable and interesting information. . . . We cannot but recommend the work to the consideration of all interested in the practical reform of our weights and measures."—*Nature.*

"For exhaustive tables of equivalent weights and measures of all sorts, and for clear demonstrations of the effects of the various systems that have been proposed or adopted, Mr. Jackson's treatise is without a rival."—*Academy.*

The Metric System and the British Standards.
A SERIES OF METRIC TABLES, in which the British Standard Measures and Weights are compared with those of the Metric System at present in Use on the Continent. By C. H. DOWLING, C.E. 8vo, 10s. 6d. strongly bound.

"Their accuracy has been certified by Professor Airy, the Astronomer-Royal."—*Builder.*

"Mr. Dowling's Tables are well put together as a ready-reckoner for the conversion of one system into the other."—*Athenæum.*

Iron and Metal Trades' Calculator.
THE IRON AND METAL TRADES' COMPANION. For expeditiously ascertaining the Value of any Goods bought or sold by Weight, from 1s. per cwt. to 112s. per cwt., and from one farthing per pound to one shilling per pound. Each Table extends from one pound to 100 tons. To which are appended Rules on Decimals, Square and Cube Root, Mensuration of Superficies and Solids, &c.; also Tables of Weights of Materials, and other Useful Memoranda. By THOS. DOWNIE. Strongly bound in leather, 396 pp., 9s.

"A most useful set of tables, and will supply a want, for nothing like them before existed."—*Building News.*

"Although specially adapted to the iron and metal trades, the tables will be found useful in every other business in which merchandise is bought and sold by weight."—*Railway News*

CROSBY LOCKWOOD & SON'S CATALOGUE.

Calculator for Numbers and Weights Combined.

THE NUMBER, WEIGHT AND FRACTIONAL CALCULATOR. Containing upwards of 250,000 Separate Calculations, showing at a glance the value at 422 different rates, ranging from $\frac{1}{16}$th of a Penny to 20s. each, or per cwt., and £20 per ton, of any number of articles consecutively, from 1 to 470.—Any number of cwts., qrs., and lbs., from 1 cwt. to 470 cwts.—Any number of tons, cwts., qrs., and lbs., from 1 to 1,000 tons. By WILLIAM CHADWICK, Public Accountant. Third Edition, Revised and Improved. 8vo, price 18s., strongly bound for Office wear and tear. [*Just published.*

*** *This work is specially adapted for the Apportionment of Mileage Charges for Railway Traffic.*

☞ *This comprehensive and entirely unique and original Calculator is adapted for the use of Accountants and Auditors, Railway Companies, Canal Companies, Shippers, Shipping Agents, General Carriers, etc.*

Ironfounders, Brassfounders, Metal Merchants, Iron Manufacturers, Ironmongers, Engineers, Machinists, Boiler Makers, Millwrights, Roofing, Bridge and Girder Makers, Colliery Proprietors, etc.

Timber Merchants, Builders, Contractors, Architects, Surveyors, Auctioneers Valuers, Brokers, Mill Owners and Manufacturers, Mill Furnishers, Merchants and General Wholesale Tradesmen.

*** OPINIONS OF THE PRESS.

"The book contains the answers to questions, and not simply a set of ingenious puzzle methods of arriving at results. It is as easy of reference for any answer or any number of answers as a dictionary, and the references are even more quickly made. For making up accounts or estimates, the book must prove invaluable to all who have any considerable quantity of calculations involving price and measure in any combination to do."—*Engineer.*

"The most perfect work of the kind yet prepared."—*Glasgow Herald.*

Comprehensive Weight Calculator.

THE WEIGHT CALCULATOR. Being a Series of Tables upon a New and Comprehensive Plan, exhibiting at One Reference the exact Value of any Weight from 1 lb. to 15 tons, at 300 Progressive Rates, from 1d. to 168s. per cwt., and containing 186,000 Direct Answers, which, with their Combinations, consisting of a single addition (mostly to be performed at sight), will afford an aggregate of 10,266,000 Answers; the whole being calculated and designed to ensure correctness and promote despatch. By HENRY HARBEN, Accountant. Fourth Edition, carefully Corrected. Royal 8vo, strongly half-bound, £1 5s.

"A practical and useful work of reference for men of business generally; it is the best of the kind we have seen.'—*Ironmonger.*

"Of priceless value to business men. It is a necessary book in all mercantile offices."—*Sheffield Independent.*

Comprehensive Discount Guide.

THE DISCOUNT GUIDE. Comprising several Series of Tables for the use of Merchants, Manufacturers, Ironmongers, and others, by which may be ascertained the exact Profit arising from any mode of using Discounts, either in the Purchase or Sale of Goods, and the method of either Altering a Rate of Discount or Advancing a Price, so as to produce, by one operation, a sum that will realise any required profit after allowing one or more Discounts: to which are added Tables of Profit or Advance from 1¼ to 90 per cent., Tables of Discount from 1¼ to 98¾ per cent., and Tables of Commission, &c., from ⅛ to 10 per cent. By HENRY HARBEN, Accountant, Author of "The Weight Calculator." New Edition, carefully Revised and Corrected. Demy 8vo, 544 pp. half-bound, £1 5s.

"A book such as this can only be appreciated by business men, to whom the saving of time means saving of money. We have the high authority of Professor J. R. Young that the tables throughout the work are constructed upon strictly accurate principles. The work is a model of typographical clearness, and must prove of great value to merchants, manufacturers, and general traders."—*British Trade Journal.*

Iron Shipbuilders' and Merchants' Weight Tables.

IRON-PLATE WEIGHT TABLES: *For Iron Shipbuilders, Engineers and Iron Merchants.* Containing the Calculated Weights of upwards of 150,000 different sizes of Iron Plates, from 1 foot by 6 in. by ¼ in. to 10 feet by 5 feet by 1 in. Worked out on the basis of 40 lbs. to the square foot of Iron of 1 inch in thickness. Carefully compiled and thoroughly Revised by H. BURLINSON and W. H. SIMPSON. Oblong 4to, 25s. half-bound.

"This work will be found of great utility. The authors have had much practical experience of what is wanting in making estimates; and the use of the book will save much time in making elaborate calculations."—*English Mechanic.*

INDUSTRIAL AND USEFUL ARTS.

Soap-making.
THE ART OF SOAP-MAKING: A Practical Handbook of the Manufacture of Hard and Soft Soaps, Toilet Soaps, etc. Including many New Processes, and a Chapter on the Recovery of Glycerine from Waste Leys. By ALEXANDER WATT, Author of "Electro-Metallurgy Practically Treated," &c. With numerous Illustrations. Fourth Edition, Revised and Enlarged. Crown 8vo, 7s. 6d. cloth. [*Just published.*

"The work will prove very useful, not merely to the technological student, but to the practical soap-boiler who wishes to understand the theory of his art."—*Chemical News.*

"Mr. Watt's book is a thoroughly practical treatise on an art which has almost no literature in our language. We congratulate the author on the success of his endeavour to fill a void in English technical literature."—*Nature.*

Paper Making.
THE ART OF PAPER MAKING: A Practical Handbook of the Manufacture of Paper from Rags, Esparto, Straw and other Fibrous Materials, Including the Manufacture of Pulp from Wood Fibre, with a Description of the Machinery and Appliances used. To which are added Details of Processes for Recovering Soda from Waste Liquors. By ALEXANDER WATT. With Illustrations. Crown 8vo, 7s. 6d. cloth. [*Just published.*

"This book is succinct, lucid, thoroughly practical, and includes everything of interest to the modern paper maker. It is the latest, most practical and most complete work on the paper-making art before the British public."—*Paper Record.*

"It may be regarded as the standard work on the subject. The book is full of valuable information. The 'Art of Paper-making,' is in every respect a model of a text-book, either for a technical class or for the private student."—*Paper and Printing Trades Journal.*

"Admirably adapted for general as well as ordinary technical reference, and as a handbook for students in technical education may be warmly commended."—*The Paper Maker's Monthly Journal.*

Leather Manufacture.
THE ART OF LEATHER MANUFACTURE. Being a Practical Handbook, in which the Operations of Tanning, Currying, and Leather Dressing are fully Described, the Principles of Tanning Explained and many Recent Processes introduced. By ALEXANDER WATT, Author of "Soap-Making," &c. With numerous Illustrations. Second Edition. Crown 8vo, 9s. cloth.

"A sound, comprehensive treatise on tanning and its accessories. This book is an eminently valuable production, which redounds to the credit of both author and publishers."—*Chemical Review.*

"This volume is technical without being tedious, comprehensive and complete without being prosy, and it bears on every page the impress of a master hand. We have never come across a better trade treatise, nor one that so thoroughly supplied an absolute want."—*Shoe and Leather Trades' Chronicle.*

Boot and Shoe Making.
THE ART OF BOOT AND SHOE-MAKING. A Practical Handbook, including Measurement, Last-Fitting, Cutting-Out, Closing and Making, with a Description of the most approved Machinery employed. By JOHN B. LENO, late Editor of *St. Crispin*, and *The Boot and Shoe-Maker*. With numerous Illustrations. Third Edition. 12mo, 2s. cloth limp.

"This excellent treatise is by far the best work ever written on the subject. A new work, embracing all modern improvements, was much wanted. This want is now satisfied. The chapter on clicking, which shows how waste may be prevented, will save fifty times the price of the book."—*Scottish Leather Trader.*

Dentistry.
MECHANICAL DENTISTRY: A Practical Treatise on the Construction of the various kinds of Artificial Dentures. Comprising also Useful Formulæ, Tables and Receipts for Gold Plate, Clasps, Solders, &c. &c. By CHARLES HUNTER. Third Edition, Revised. With upwards of 100 Wood Engravings. Crown 8vo, 3s. 6d. cloth.

"The work is very practical."—*Monthly Review of Dental Surgery.*

"We can strongly recommend Mr. Hunter's treatise to all students preparing for the profession of dentistry, as well as to every mechanical dentist."—*Dublin Journal of Medical Science.*

Wood Engraving.
WOOD ENGRAVING: A Practical and Easy Introduction to the Study of the Art. By WILLIAM NORMAN BROWN. Second Edition. With numerous Illustrations. 12mo, 1s. 6d. cloth limp.

"The book is clear and complete, and will be useful to anyone wanting to understand the first elements of the beautiful art of wood engraving."—*Graphic.*

HANDYBOOKS FOR HANDICRAFTS. By PAUL N. HASLUCK.

Metal Turning.
THE METAL TURNER'S HANDYBOOK. A Practical Manual for Workers at the Foot-Lathe: Embracing Information on the Tools, Appliances and Processes employed in Metal Turning. By PAUL N. HASLUCK, Author of "Lathe-Work." With upwards of One Hundred Illustrations. Second Edition, Revised. Crown 8vo, 2s. cloth.
"Clearly and concisely written, excellent in every way."—*Mechanical World.*

Wood Turning.
THE WOOD TURNER'S HANDYBOOK. A Practical Manual for Workers at the Lathe: Embracing Information on the Tools, Appliances and Processes Employed in Wood Turning. By PAUL N. HASLUCK. With upwards of One Hundred Illustrations. Crown 8vo, 2s. cloth.
"We recommend the book to young turners and amateurs. A multitude of workmen have hitherto sought in vain for a manual of this special industry."—*Mechanical World.*

WOOD AND METAL TURNING. By P. N. HASLUCK. (Being the Two preceding Vols. bound together.) 300 pp., with upwards of 200 Illustrations, crown 8vo, 3s. 6d. cloth.

Watch Repairing.
THE WATCH JOBBER'S HANDYBOOK. A Practical Manual on Cleaning, Repairing and Adjusting. Embracing Information on the Tools, Materials, Appliances and Processes Employed in Watchwork. By PAUL N. HASLUCK. With upwards of One Hundred Illustrations. Cr. 8vo, 2s. cloth.
"All young persons connected with the trade should acquire and study this excellent, and at the same time, inexpensive work."—*Clerkenwell Chronicle.*

Clock Repairing.
THE CLOCK JOBBER'S HANDYBOOK: A Practical Manual on Cleaning, Repairing and Adjusting. Embracing Information on the Tools, Materials, Appliances and Processes Employed in Clockwork. By PAUL N. HASLUCK. With upwards of 100 Illustrations. Cr. 8vo, 2s. cloth.
"Of inestimable service to those commencing the trade."—*Coventry Standard.*

WATCH AND CLOCK JOBBING. By P. N. HASLUCK. (Being the Two preceding Vols. bound together.) 320 pp., with upwards of 200 Illustrations, crown 8vo, 3s. 6d. cloth.

Pattern Making.
THE PATTERN MAKER'S HANDYBOOK. A Practical Manual, embracing Information on the Tools, Materials and Appliances employed in Constructing Patterns for Founders. By PAUL N. HASLUCK. With One Hundred Illustrations. Crown 8vo, 2s. cloth.
"This handy volume contains sound information of considerable value to students and artificers."—*Hardware Trades Journal.*

Mechanical Manipulation.
THE MECHANIC'S WORKSHOP HANDYBOOK. A Practical Manual on Mechanical Manipulation. Embracing Information on various Handicraft Processes, with Useful Notes and Miscellaneous Memoranda. By PAUL N. HASLUCK. Crown 8vo, 2s. cloth.
"It is a book which should be found in every workshop, as it is one which will be continually referred to for a very great amount of standard information."—*Saturday Review.*

Model Engineering.
THE MODEL ENGINEER'S HANDYBOOK: A Practical Manual on Model Steam Engines. Embracing Information on the Tools, Materials and Processes Employed in their Construction. By PAUL N. HASLUCK. With upwards of 100 Illustrations. Crown 8vo, 2s. cloth.
"By carefully going through the work, amateurs may pick up an excellent notion of the construction of full-sized steam engines."—*Telegraphic Journal.*

Cabinet Making.
THE CABINET WORKER'S HANDYBOOK: A Practical Manual, embracing Information on the Tools, Materials, Appliances and Processes employed in Cabinet Work. By PAUL N. HASLUCK, Author of "Lathe Work," &c. With upwards of 100 Illustrations. Crown 8vo, 2s. cloth. [*Just published.*
"Thoroughly practical throughout. The amateur worker in wood will find it most useful."—*Glasgow Herald.*

INDUSTRIAL AND USEFUL ARTS. 33

Electrolysis of Gold, Silver, Copper, etc.
ELECTRO-DEPOSITION : A Practical Treatise on the Electrolysis of Gold, Silver, Copper, Nickel, and other Metals and Alloys. With descriptions of Voltaic Batteries, Magneto and Dynamo-Electric Machines, Thermopiles, and of the Materials and Processes used in every Department of the Art, and several Chapters on Electro-Metallurgy. By ALEXANDER WATT. Third Edition, Revised and Corrected. Crown 8vo, 9s. cloth.
"Eminently a book for the practical worker in electro-deposition. It contains practical descriptions of methods, processes and materials as actually pursued and used in the workshop."—*Engineer.*

Electro-Metallurgy.
ELECTRO-METALLURGY; Practically Treated. By ALEXANDER WATT, Author of "Electro-Deposition," &c. Ninth Edition, Enlarged and Revised, with Additional Illustrations, and including the most recent Processes. 12mo, 4s. cloth boards.
"From this book both amateur and artisan may learn everything necessary for the successful prosecution of electroplating."—*Iron.*

Electroplating.
ELECTROPLATING: A Practical Handbook on the Deposition of Copper, Silver, Nickel, Gold, Aluminium, Brass, Platinum, &c. &c. With Descriptions of the Chemicals, Materials, Batteries and Dynamo Machines used in the Art. By J. W. URQUHART, C.E. Second Edition, with Additions. Numerous Illustrations. Crown 8vo, 5s. cloth.
"An excellent practical manual."—*Engineering.*
"An excellent work, giving the newest information."—*Horological Journal.*

Electrotyping.
ELECTROTYPING: The Reproduction and Multiplication of Printing Surfaces and Works of Art by the Electro-deposition of Metals. By J. W. URQUHART, C.E. Crown 8vo, 5s. cloth.
"The book is thoroughly practical. The reader is, therefore, conducted through the leading aws of electricity, then through the metals used by electrotypers, the apparatus, and the depositing processes, up to the final preparation of the work."—*Art Journal.*

Horology.
A TREATISE ON MODERN HOROLOGY, in Theory and Practice. Translated from the French of CLAUDIUS SAUNIER, by JULIEN TRIPPLIN, F.R.A.S., and EDWARD RIGG, M.A., Assayer in the Royal Mint. With 78 Woodcuts and 22 Coloured Plates. Second Edition. Royal 8vo, £2 2s. cloth ; £2 10s. half-calf.
"There is no horological work in the English language at all to be compared to this production of M. Saunier's for clearness and completeness. It is alike good as a guide for the student and as a re'erence for the experienced horologist and skilled workman."—*Horological Journal.*
"The latest, the most complete, and the most reliable of those literary productions to which continental watchmakers are indebted for the mechanical superiority over their English brethren—in fact, the Book of Books, is M. Saunier's 'Treatise.'"—*Watchmaker, Jeweller and Silversmith.*

Watchmaking.
THE WATCHMAKER'S HANDBOOK. A Workshop Companion for those engaged in Watchmaking and the Allied Mechanical Arts. From the French of CLAUDIUS SAUNIER. Enlarged by JULIEN TRIPPLIN, F.R.A.S., and EDWARD RIGG, M.A., Assayer in the Royal Mint. Woodcuts and Copper Plates. Third Edition, Revised. Crown 8vo, 9s. cloth.
"Each part is truly a treatise in itself. The arrangement is good and the language is clear and concise. It is an admirable guide for the young watchmaker."—*Engineering.*
"It is impossible to speak too highly of its excellence. It fulfils every requirement in a handbook intended for the use of a workman."—*Watch and Clockmaker.*
"This book contains an immense number of practical details bearing on the daily occupation of a watchmaker."—*Watchmaker and Metalworker* (Chicago).

Goldsmiths' Work.
THE GOLDSMITH'S HANDBOOK. By GEORGE E. GEE, Jeweller, &c. Third Edition, considerably Enlarged. 12mo, 3s. 6d. cl. bds.
"A good, sound educator, and will be accepted as an authority."—*Horological Journal.*

Silversmiths' Work.
THE SILVERSMITH'S HANDBOOK. By GEORGE E. GEE, Jeweller, &c. Second Edition, Revised, with numerous Illustrations. 12mo, 3s. 6d. cloth boards.
"Workers in the trade will speedily discover its merits when they sit down to study it."—*English Mechanic.*

∗ *The above two works together, strongly half-bound, price 7s.*

Bread and Biscuit Baking.

THE BREAD AND BISCUIT BAKER'S AND SUGAR-BOILER'S ASSISTANT. Including a large variety of Modern Recipes. With Remarks on the Art of Bread-making. By ROBERT WELLS, Practical Baker. Second Edition, with Additional Recipes. Crown 8vo, 2s. cloth.
[*Just published.*
"A large number of wrinkles for the ordinary cook, as well as the baker."—*Saturday Review.*

Confectionery.

THE PASTRYCOOK AND CONFECTIONER'S GUIDE. For Hotels, Restaurants and the Trade in general, adapted also for Family Use. By ROBERT WELLS, Author of "The Bread and Biscuit Baker's and Sugar Boiler's Assistant." Crown 8vo, 2s. cloth. [*Just published.*
"We cannot speak too highly of this really excellent work. In these days of keen competition our readers cannot do better than purchase this book."—*Bakers' Times.*

Ornamental Confectionery.

ORNAMENTAL CONFECTIONERY: A Guide for Bakers, Confectioners and Pastrycooks; including a variety of Modern Recipes, and Remarks on Decorative and Coloured Work. With 129 Original Designs. By ROBERT WELLS. Crown 8vo, 5s. cloth.
"A valuable work, practical, and should be in the hands of every baker and confectioner. The illustrative designs are alone worth treble the amount charged for the whole work."—*Bakers' Times.*

Flour Confectionery.

THE MODERN FLOUR CONFECTIONER. Wholesale and Retail. Containing a large Collection of Recipes for Cheap Cakes, Biscuits, &c. With Remarks on the Ingredients used in their Manufacture, &c. By R. WELLS, Author of "Ornamental Confectionery," "The Bread and Biscuit Baker," "The Pastrycook's Guide," &c. Crown 8vo, 2s. cloth.
[*Just published.*

Laundry Work.

LAUNDRY MANAGEMENT. A Handbook for Use in Private and Public Laundries, Including Descriptive Accounts of Modern Machinery and Appliances for Laundry Work. By the EDITOR of "The Laundry Journal." With numerous Illustrations. Crown 8vo, 2s. 6d. cloth.

CHEMICAL MANUFACTURES & COMMERCE.

Alkali Trade, Manufacture of Sulphuric Acid, etc.

A MANUAL OF THE ALKALI TRADE, including the Manufacture of Sulphuric Acid, Sulphate of Soda, and Bleaching Powder. By JOHN LOMAS. 390 pages. With 232 Illustrations and Working Drawings. Second Edition. Royal 8vo, £1 10s. cloth.
"This book is written by a manufacturer for manufacturers. The working details of the most approved forms of apparatus are given, and these are accompanied by no less than 232 wood engravings, all of which may be used for the purposes of construction. Every step in the manufacture is very fully described in this manual, and each improvement explained."—*Athenæum.*

The Blowpipe.

THE BLOWPIPE IN CHEMISTRY, MINERALOGY, AND GEOLOGY. Containing all known Methods of Anhydrous Analysis, Working Examples, and Instructions for Making Apparatus. By Lieut.-Col. W. A. Ross, R.A. With 120 Illustrations. New Edition. Crown 8vo, 5s.
"The student who goes through the course of experimentation here laid down will gain a better insight into inorganic chemistry and mineralogy than if he had 'got up' any of the best text-books of the day, and passed any number of examinations in their contents."—*Chemical News.*

Commercial Chemical Analysis.

THE COMMERCIAL HANDBOOK OF CHEMICAL ANALYSIS; or, Practical Instructions for the determination of the Intrinsic or Commercial Value of Substances used in Manufactures, Trades, and the Arts. By A. NORMANDY. New Edition by H. M. NOAD, F.R.S. Cr. 8vo, 12s. 6d. cl.
"Essential to the analysts appointed under the new Act. The most recent results are given, and the work is well edited and carefully written."—*Nature.*

AGRICULTURE, FARMING, GARDENING, etc.

Brewing.
A HANDBOOK FOR YOUNG BREWERS. By HERBERT EDWARDS WRIGHT, B.A. An Entirely New Edition, much Enlarged.
[*In the press.*

Analysis and Valuation of Fuels.
FUELS: SOLID, LIQUID AND GASEOUS, Their Analysis and Valuation. For the Use of Chemists and Engineers. By H. J. PHILLIPS, F.C.S., Analytical and Consulting Chemist to the Great Eastern Railway. Crown 8vo, 3s. 6d. cloth.   [*Just published*
"Ought to have its place in the laboratory of every metallurgical establishment, and wherever fuel is used on a large scale."—*Chemical News.*
"Mr. Phillips' new book cannot fail to be of wide interest, especially at the present time."—*Railway News.*

Dye-Wares and Colours.
THE MANUAL OF COLOURS AND DYE-WARES: Their Properties, Applications, Valuation, Impurities, and Sophistications. For the use of Dyers, Printers, Drysalters, Brokers, &c. By J. W. SLATER. Second Edition, Revised and greatly Enlarged. Crown 8vo, 7s. 6d. cloth.
"A complete encyclopædia of the *materia tinctoria*. The information given respecting each article is full and precise, and the methods of determining the value of articles such as these, so liable to sophistication, are given with clearness, and are practical as well as valuable."—*Chemist and Druggist.*
"There is no other work which covers precisely the same ground. To students preparing or examinations in dyeing and printing it will prove exceedingly useful."—*Chemical News.*

Pigments.
THE ARTIST'S MANUAL OF PIGMENTS. Showing their Composition, Conditions of Permanency, Non-Permanency, and Adulterations; Effects in Combination with Each Other and with Vehicles; and the most Reliable Tests of Purity. Together with the Science and Arts Department's Examination Questions on Painting. By H. C. STANDAGE. Second Edition. Crown 8vo, 2s. 6d. cloth.
"This work is indeed *multum-in-parvo*, and we can, with good conscience, recommend it to all who come in contact with pigments, whether as makers, dealers or users."—*Chemical Review.*

Gauging. Tables and Rules for Revenue Officers, Brewers, etc.
A POCKET BOOK OF MENSURATION AND GAUGING: Containing Tables, Rules and Memoranda for Revenue Officers, Brewers, Spirit Merchants, &c. By J. B. MANT (Inland Revenue). Second Edition, Revised. Oblong 18mo, 4s. leather, with elastic band.   [*Just published.*
"This handy and useful book is adapted to the requirements of the Inland Revenue Department, and will be a favourite book of reference. The range of subjects is comprehensive, and the arrangement simple and clear."—*Civilian.*
"Should be in the hands of every practical brewer."—*Brewers' Journal.*

AGRICULTURE, FARMING, GARDENING, etc.

Youatt and Burn's Complete Grazier.
THE COMPLETE GRAZIER, and FARMER'S and CATTLE-BREEDER'S ASSISTANT. A Compendium of Husbandry; especially in the departments connected with the Breeding, Rearing, Feeding, and General Management of Stock; the Management of the Dairy, &c. With Directions for the Culture and Management of Grass Land, of Grain and Root Crops, the Arrangement of Farm Offices, the use of Implements and Machines, and on Draining, Irrigation, Warping, &c.; and the Application and Relative Value of Manures. By WILLIAM YOUATT, Esq., V.S., and ROBERT SCOTT BURN. A New Edition, partly Re-Written and greatly Enlarged by W. FREAM, B.Sc. Lond., LL.D. One large 8vo Volume, nearly 1,000 pages.
[*In preparation.*

Agricultural Facts and Figures.
NOTE-BOOK OF AGRICULTURAL FACTS AND FIGURES FOR FARMERS AND FARM STUDENTS. By PRIMROSE MCCONNELL, Fellow of the Highland and Agricultural Society; late Professor of Agriculture, Glasgow Veterinary College. Third Edition. Royal 32mo, full roan, gilt edges, with elastic band, 4s.
"The most complete and comprehensive Note-book for Farmers and Farm Students that we have seen. It literally teems with information, and we can cordially recommend it to all connected with agriculture."—*North British Agriculturist.*

Flour Manufacture, Milling, etc.

FLOUR MANUFACTURE: A Treatise on Milling Science and Practice. By FRIEDRICH KICK, Imperial Regierungsrath, Professor of Mechanical Technology in the Imperial German Polytechnic Institute, Prague. Translated from the Second Enlarged and Revised Edition with Supplement. By H. H. P. POWLES, A.M.I.C.E. Nearly 400 pp. Illustrated with 28 Folding Plates, and 167 Woodcuts. Royal 8vo, 25s. cloth.

"This valuable work is, and will remain, the standard authority on the science of milling. . . . The miller who has read and digested this work will have laid the foundation, so to speak, of a successful career; he will have acquired a number of general principles which he can proceed to apply. In this handsome volume we at last have the accepted text-book of modern milling in good, sound English, which has little, if any, trace of the German idiom."—*The Miller.*
"The appearance of this celebrated work in English is very opportune, and British millers will, we are sure, not be slow in availing themselves of its pages."—*Millers' Gazette.*

Small Farming.

SYSTEMATIC SMALL FARMING; or, The Lessons of my Farm. Being an Introduction to Modern Farm Practice for Small Farmers in the Culture of Crops; The Feeding of Cattle; The Management of the Dairy, Poultry and Pigs, &c. &c. By ROBERT SCOTT BURN, Author of "Outlines of Landed Estates' Management." Numerous Illusts., cr. 8vo, 6s. cloth.

"This is the completest book of its class we have seen, and one which every amateur farmer will read with pleasure and accept as a guide."—*Field.*
"The volume contains a vast amount of useful information. No branch of farming is left untouched, from the labour to be done to the results achieved. It may be safely recommended to all who think they will be in paradise when they buy or rent a three-acre farm."—*Glasgow Herald.*

Modern Farming.

OUTLINES OF MODERN FARMING. By R. SCOTT BURN. Soils, Manures, and Crops—Farming and Farming Economy—Cattle, Sheep, and Horses — Management of Dairy, Pigs and Poultry — Utilisation of Town-Sewage, Irrigation, &c. Sixth Edition. In One Vol., 1,250 pp., half-bound, profusely Illustrated, 12s.

"The aim of the author has been to make his work at once comprehensive and trustworthy, and in this aim he has succeeded to a degree which entitles him to much credit."—*Morning Advertiser.*   "No farmer should be without this book."—*Banbury Guardian.*

Agricultural Engineering.

FARM ENGINEERING, THE COMPLETE TEXT-BOOK OF. Comprising Draining and Embanking; Irrigation and Water Supply; Farm Roads, Fences, and Gates; Farm Buildings, their Arrangement and Construction, with Plans and Estimates; Barn Implements and Machines; Field Implements and Machines; Agricultural Surveying, Levelling, &c. By Prof. JOHN SCOTT, Editor of the "Farmers' Gazette," late Professor of Agriculture and Rural Economy at the Royal Agricultural College, Cirencester, &c. &c. In One Vol., 1,150 pages, half-bound, with over 600 Illustrations, 12s.

"Written with great care, as well as with knowledge and ability. The author has done his work well; we have found him a very trustworthy guide wherever we have tested his statements. The volume will be of great value to agricultural students."—*Mark Lane Express.*
"For a young agriculturist we know of no handy volume likely to be more usefully studied."—*Bell's Weekly Messenger.*

English Agriculture.

THE FIELDS OF GREAT BRITAIN: A Text-Book of Agriculture, adapted to the Syllabus of the Science and Art Department. For Elementary and Advanced Students. By HUGH CLEMENTS (Board of Trade). Second Ed., Revised, with Additions. 18mo, 2s. 6d. cl. [*Just published.*

"A most comprehensive volume, giving a mass of information."—*Agricultural Economist.*
"It is a long time since we have seen a book which has pleased us more, or which contains such a vast and useful fund of knowledge."—*Educational Times.*

Tables for Farmers, etc.

TABLES, MEMORANDA, AND CALCULATED RESULTS for Farmers, Graziers, Agricultural Students, Surveyors, Land Agents Auctioneers, etc. With a New System of Farm Book-keeping. Selected and Arranged by SIDNEY FRANCIS. Second Edition, Revised. 272 pp., waistcoat-pocket size, 1s. 6d. limp leather. [*Just published.*

"Weighing less than 1 oz., and occupying no more space than a match box, it contains a mass of facts and calculations which has never before, in such handy form, been obtainable. Every operation on the farm is dealt with. The work may be taken as thoroughly accurate, the whole of the tables having been revised by Dr. Fream. We cordially recommend it."—*Bell's Weekly Messenger.*
"A marvellous little book. . . . The agriculturist who possesses himself of it will not be disappointed with his investment."—*The Farm.*

AGRICULTURE, FARMING, GARDENING, etc. 37

Farm and Estate Book-keeping.
BOOK-KEEPING FOR FARMERS & ESTATE OWNERS. A Practical Treatise, presenting, in Three Plans, a System adapted for all Classes of Farms. By JOHNSON M. WOODMAN, Chartered Accountant. Second Edition, Revised. Cr. 8vo, 3s. 6d. cl. bds.; or 2s. 6d. cl. limp. [*Just published.*
"The volume is a capital study of a most important subject."—*Agricultural Gazette.*
"Will be found of great assistance by those who intend to commence a system of book-keeping, the author's examples being clear and explicit, and his explanations, while full and accurate, being to a large extent free from technicalities."—*Live Stock Journal.*

Farm Account Book.
WOODMAN'S YEARLY FARM ACCOUNT BOOK. Giving a Weekly Labour Account and Diary, and showing the Income and Expenditure under each Department of Crops, Live Stock, Dairy, &c. &c. With Valuation, Profit and Loss Account, and Balance Sheet at the end of the Year, and an Appendix of Forms. Ruled and Headed for Entering a Complete Record of the Farming Operations. By JOHNSON M. WOODMAN, Chartered Accountant, Author of "Book-keeping for Farmers." Folio, 7s. 6d. half bound. [*culture.*
"Contains every requisite form for keeping farm accounts readily and accurately."—*Agri-*

Early Fruits, Flowers and Vegetables.
THE FORCING GARDEN; or, How to Grow Early Fruits, Flowers, and Vegetables. With Plans and Estimates for Building Glasshouses, Pits and Frames. Containing also Original Plans for Double Glazing, a New Method of Growing the Gooseberry under Glass, &c. &c., and on Ventilation, Protecting Vine Borders, &c. With Illustrations. By SAMUEL WOOD. Crown 8vo, 3s. 6d. cloth.
"A good book, and fairly fills a place that was in some degree vacant. The book is written with great care, and contains a great deal of valuable teaching."—*Gardeners' Magazine.*
"Mr. Wood's book is an original and exhaustive answer to the question 'How to Grow Early Fruits, Flowers and Vegetables!'"—*Land and Water.*

Good Gardening.
A PLAIN GUIDE TO GOOD GARDENING; or, How to Grow Vegetables, Fruits, and Flowers. With Practical Notes on Soils, Manures, Seeds, Planting, Laying-out of Gardens and Grounds, &c. By S. WOOD. Fourth Edition, with considerable Additions, &c., and numerous Illustrations. Crown 8vo, 3s. 6d. cloth.
"A very good book, and one to be highly recommended as a practical guide. The practical directions are excellent."—*Athenæum.*
"May be recommended to young gardeners, cottagers, and specially to amateurs, for the plain, simple, and trustworthy information it gives on common matters too often neglected."—*Gardeners' Chronicle.*

Gainful Gardening.
MULTUM-IN-PARVO GARDENING; cr, How to make One Acre of Land produce £620 a-year by the Cultivation of Fruits and Vegetables; also, How to Grow Flowers in Three Glass Houses, so as to realise £176 per annum clear Profit. By SAMUEL WOOD, Author of "Good Gardening," &c. Fifth and cheaper Edition, Revised, with Additions. Crown 8vo, 1s. sewed.
"We are bound to recommend it as not only suited to the case of the amateur and gentleman's gardener, but to the market grower."—*Gardeners' Magazine.*

Gardening for Ladies.
THE LADIES' MULTUM-IN-PARVO FLOWER GARDEN, and Amateurs' Complete Guide. By S. WOOD. With Illusts. Cr. 8vo, 3s. 6d. cl.
"This volume contains a good deal of sound, common sense instruction."—*Florist.*
"Full of shrewd hints and useful instructions, based on a lifetime of experience."—*Scotsman.*

Receipts for Gardeners.
GARDEN RECEIPTS. Edited by CHARLES W. QUIN. 12mo, 1s. 6d. cloth limp.
"A useful and handy book, containing a good deal of valuable information."—*Athenæum.*

Market Gardening.
MARKET AND KITCHEN GARDENING. By Contributors to "The Garden." Compiled by C. W. SHAW, late Editor of "Gardening Illustrated." 12mo, 3s. 6d. cloth boards. [*Just published.*
"The most valuable compendium of kitchen and market-garden work published."—*Farmer.*

Cottage Gardening.
COTTAGE GARDENING; or, Flowers, Fruits, and Vegetables for Small Gardens. By E. HOBDAY. 12mo, 1s. 6d. cloth limp.
"Contains much useful information at a small charge."—*Glasgow Herald.*

LAND AND ESTATE MANAGEMENT, LAW, etc.

Hudson's Land Valuer's Pocket-Book.

THE LAND VALUER'S BEST ASSISTANT: Being Tables on a very much Improved Plan, for Calculating the Value of Estates. With Tables for reducing Scotch, Irish, and Provincial Customary Acres to Statute Measure, &c. By R. HUDSON, C.E. New Edition. Royal 32mo, leather, elastic band, 4s.

"This new edition includes tables for ascertaining the value of leases for any term of years; and for showing how to lay out plots of ground of certain acres in forms, square, round, &c., with valuable rules for ascertaining the probable worth of standing timber to any amount; and is of incalculable value to the country gentleman and professional man."—*Farmers' Journal.*

Ewart's Land Improver's Pocket-Book.

THE LAND IMPROVER'S POCKET-BOOK OF FORMULÆ, TABLES and MEMORANDA *required in any Computation relating to the Permanent Improvement of Landed Property.* By JOHN EWART, Land Surveyor and Agricultural Engineer. Second Edition, Revised. Royal 32mo, oblong, leather, gilt edges, with elastic band, 4s.

"A compendious and handy little volume."—*Spectator.*

Complete Agricultural Surveyor's Pocket-Book.

THE LAND VALUER'S AND LAND IMPROVER'S COMPLETE POCKET-BOOK. Consisting of the above Two Works bound together. Leather, gilt edges, with strap, 7s. 6d.

"Hudson's book is the best ready-reckoner on matters relating to the valuation of land and crops, and its combination with Mr. Ewart's work greatly enhances the value and usefulness of the latter-mentioned. . . . It is most useful as a manual for reference."—*North of England Farmer.*

Auctioneer's Assistant.

THE APPRAISER, AUCTIONEER, BROKER, HOUSE AND ESTATE AGENT AND VALUER'S POCKET ASSISTANT, for the Valuation for Purchase, Sale, or Renewal of Leases, Annuities and Reversions, and of property generally; with Prices for Inventories, &c. By JOHN WHEELER, Valuer, &c. Fifth Edition, re-written and greatly extended by C. NORRIS, Surveyor, Valuer, &c. Royal 32mo, 5s. cloth.

"A neat and concise book of reference, containing an admirable and clearly-arranged list of prices for inventories, and a very practical guide to determine the value of furniture, &c."—*Standard.*

"Contains a large quantity of varied and useful information as to the valuation for purchase, sale, or renewal of leases, annuities and reversions, and of property generally, with prices for inventories, and a guide to determine the value of interior fittings and other effects."—*Builder.*

Auctioneering.

AUCTIONEERS: *Their Duties and Liabilities.* A Manual of Instruction and Counsel for the Young Auctioneer. By ROBERT SQUIBBS, Auctioneer. Second Edition, Revised and partly Re-written. Demy 8vo, 12s. 6d. cloth. [*Just published.*

"The position and duties of auctioneers treated compendiously and clearly."—*Builder.*

"Every auctioneer ought to possess a copy of this excellent work."—*Ironmonger.*

"Of great value to the profession. . . . We readily welcome this book from the fact that it treats the subject in a manner somewhat new to the profession."—*Estates Gazette.*

Legal Guide for Pawnbrokers.

THE PAWNBROKERS', FACTORS' AND MERCHANTS' GUIDE TO THE LAW OF LOANS AND PLEDGES. With the Statutes and a Digest of Cases on Rights and Liabilities, Civil and Criminal, as to Loans and Pledges of Goods, Debentures, Mercantile and other Securities. By H. C. FOLKARD, Esq., Barrister-at-Law, Author of "The Law of Slander and Libel," &c. With Additions and Corrections. Fcap. 8vo, 3s. 6d. cloth.

"This work contains simply everything that requires to be known concerning the department of the law of which it treats. We can safely commend the book as unique and very nearly perfect."—*Iron.*

"The task undertaken by Mr. Folkard has been very satisfactorily performed. . . . Such explanations as are needful have been supplied with great clearness and with due regard to brevity."—*City Press.*

LAND AND ESTATE MANAGEMENT, LAW, etc. 39

Law of Patents.
PATENTS FOR INVENTIONS, AND HOW TO PROCURE THEM. Compiled for the Use of Inventors, Patentees and others. By G. G. M. HARDINGHAM, Assoc.Mem.Inst.C.E., &c. Demy 8vo, cloth, price 2s. 6d. [*Just published.*

Metropolitan Rating Appeals.
REPORTS OF APPEALS HEARD BEFORE THE COURT OF GENERAL ASSESSMENT SESSIONS, from the Year 1871 to 1885. By EDWARD RYDE and ARTHUR LYON RYDE. Fourth Edition, brought down to the Present Date, with an Introduction to the Valuation (Metropolis) Act, 1869, and an Appendix by WALTER C. RYDE, of the Inner Temple, Barrister-at-Law. 8vo, 16s. cloth.

"A useful work, occupying a place mid-way between a handbook for a lawyer and a guide to the surveyor It is compiled by a gentleman eminent in his profession as a land agent, whose specialty, it is acknowledged, lies in the direction of assessing property for rating purposes."—*Land Agents' Record.*
"It is an indispensable work of reference for all engaged in assessment business."—*Journal of Gas Lighting.*

House Property.
HANDBOOK OF HOUSE PROPERTY. A Popular and Practical Guide to the Purchase, Mortgage, Tenancy, and Compulsory Sale of Houses and Land, including the Law of Dilapidations and Fixtures; with Examples of all kinds of Valuations, Useful Information on Building, and Suggestive Elucidations of Fine Art. By E. L. TARBUCK, Architect and Surveyor. Fourth Edition, Enlarged. 12mo, 5s. cloth.
"The advice is thoroughly practical."—*Law Journal.*
"For all who have dealings with house property, this is an indispensable guide."—*Decoration.*
"Carefully brought up to date, and much improved by the addition of a division on fine art."
"A well written and thoughtful work."—*Land Agent's Record.*

Inwood's Estate Tables.
TABLES FOR THE PURCHASING OF ESTATES, Freehold, Copyhold, or Leasehold; Annuities, Advowsons, etc., and for the Renewing of Leases held under Cathedral Churches, Colleges, or other Corporate bodies, for Terms of Years certain, and for Lives; also for Valuing Reversionary Estates, Deferred Annuities, Next Presentations, &c.; together with SMART's Five Tables of Compound Interest, and an Extension of the same to Lower and Intermediate Rates. By W. INWOOD. 23rd Edition, with considerable Additions, and new and valuable Tables of Logarithms for the more Difficult Computations of the Interest of Money, Discount, Annuities, &c., by M. FEDOR THOMAN, of the Société Crédit Mobilier of Paris. Crown 8vo, 8s. cloth.

"Those interested in the purchase and sale of estates, and in the adjustment of compensation cases, as well as in transactions in annuities, life insurances, &c., will find the present edition of eminent service."—*Engineering.*
"'Inwood's Tables' still maintain a most enviable reputation. The new issue has been enriched by large additional contributions by M. Fedor Thoman, whose carefully arranged Tables cannot fail to be of the utmost utility."—*Mining Journal.*

Agricultural and Tenant-Right Valuation.
THE AGRICULTURAL AND TENANT-RIGHT-VALUER'S ASSISTANT. A Practical Handbook on Measuring and Estimating the Contents, Weights and Values of Agricultural Produce and Timber, the Values of Estates and Agricultural Labour, Forms of Tenant-Right-Valuations, Scales of Compensation under the Agricultural Holdings Act, 1883, &c. &c. By TOM BRIGHT, Agricultural Surveyor. Crown 8vo, 3s. 6d. cloth.
"Full of tables and examples in connection with the valuation of tenant-right, estates, labour, contents, and weights of timber, and farm produce of all kinds."—*Agricultural Gazette.*
"An eminently practical handbook, full of practical tables and data of undoubted interest and value to surveyors and auctioneers in preparing valuations of all kinds."—*Farmer.*

Plantations and Underwoods.
POLE PLANTATIONS AND UNDERWOODS: A Practical Handbook on Estimating the Cost of Forming, Renovating, Improving and Grubbing Plantations and Underwoods, their Valuation for Purposes of Transfer, Rental, Sale or Assessment. By TOM BRIGHT, F.S.Sc., Author of "The Agricultural and Tenant-Right-Valuer's Assistant," &c. Crown 8vo, 3s. 6d. cloth. [*Just published.*
"Will be found very useful to those who are actually engaged in managing wood."—*Bell's Weekly Messenger.*
"To valuers, foresters and agents it will be a welcome aid."—*North British Agriculturist.*
"Well calculated to assist the valuer in the discharge of his duties, and of undoubted interest and use both to surveyors and auctioneers in preparing valuations of all kinds."—*Kent Herald.*

A Complete Epitome of the Laws of this Country.

EVERY MAN'S OWN LAWYER: A Handy-Book of the Principles of Law and Equity. By A BARRISTER. Twenty-eighth Edition. Revised and Enlarged. Including the Legislation of 1890, and including careful digests of *The Bankruptcy Act*, 1890; the *Directors' Liability Act*, 1890; the *Partnership Act*, 1890; the *Intestates' Estates Act*, 1890; the *Settled Land Act*, 1890; the *Housing of the Working Classes Act*, 1890; the *Infectious Disease (Prevention) Act*, 1890; the *Allotments Act*, 1890; the *Tenants' Compensation Act*, 1890; and the *Trustees' Appointment Act*, 1890; while other new Acts have been duly noted. Crown 8vo, 688 pp., price 6s. 8d. (saved at every consultation!), strongly bound in cloth. [*Just published.*

*** THE BOOK WILL BE FOUND TO COMPRISE (AMONGST OTHER MATTER)—

THE RIGHTS AND WRONGS OF INDIVIDUALS—LANDLORD AND TENANT—VENDORS AND PURCHASERS—PARTNERS AND AGENTS—COMPANIES AND ASSOCIATIONS—MASTERS, SERVANTS AND WORKMEN—LEASES AND MORTGAGES—CHURCH AND CLERGY, RITUAL—LIBEL AND SLANDER—CONTRACTS AND AGREEMENTS—BONDS AND BILLS OF SALE—CHEQUES, BILLS AND NOTES—RAILWAY AND SHIPPING LAW—BANKRUPTCY AND INSURANCE—BORROWERS, LENDERS AND SURETIES—CRIMINAL LAW—PARLIAMENTARY ELECTIONS—COUNTY COUNCILS—MUNICIPAL CORPORATIONS—PARISH LAW, CHURCHWARDENS, ETC.—INSANITARY DWELLINGS AND AREAS—PUBLIC HEALTH AND NUISANCES—FRIENDLY AND BUILDING SOCIETIES—COPYRIGHT AND PATENTS—TRADE MARKS AND DESIGNS—HUSBAND AND WIFE, DIVORCE, ETC.—TRUSTEES AND EXECUTORS—GUARDIAN AND WARD, INFANTS, ETC.—GAME LAWS AND SPORTING—HORSES, HORSE-DEALING AND DOGS—INNKEEPERS, LICENSING, ETC.—FORMS OF WILLS, AGREEMENTS, ETC. ETC.

NOTE.—*The object of this work is to enable those who consult it to help themselves to the law; and thereby to dispense, as far as possible, with professional assistance and advice. There are many wrongs and grievances which persons submit to from time to time through not knowing how or where to apply for redress; and many persons have as great a dread of a lawyer's office as of a lion's den. With this book at hand it is believed that many a SIX-AND-EIGHTPENCE may be saved; many a wrong redressed; many a right reclaimed; many a law suit avoided; and many an evil abated. The work has established itself as the standard legal adviser of all classes, and also made a reputation for itself as a useful book of reference for lawyers residing at a distance from law libraries, who are glad to have at hand a work embodying recent decisions and enactments.*

*** OPINIONS OF THE PRESS.

"It is a complete code of English Law, written in plain language, which all can understand. . . Should be in the hands of every business man, and all who wish to abolish lawyers' bills."—*Weekly Times.*

"A useful and concise epitome of the law, compiled with considerable care."—*Law Magazine.*

"A complete digest of the most useful facts which constitute English law."—*Globe.*

"This excellent handbook. . . . Admirably done, admirably arranged, and admirably cheap."—*Leeds Mercury.*

"A concise, cheap and complete epitome of the English law. So plainly written that he who runs may read, and he who reads may understand."—*Figaro.*

"A dictionary of legal facts well put together. The book is a very useful one."—*Spectator.*

"A work which has long been wanted, which is thoroughly well done, and which we most cordially recommend."—*Sunday Times.*

"The latest edition of this popular book ought to be in every business establishment, and on every library table."—*Sheffield Post.*

Private Bill Legislation and Provisional Orders.

HANDBOOK FOR THE USE OF SOLICITORS AND ENGINEERS Engaged in Promoting Private Acts of Parliament and Provisional Orders, for the Authorization of Railways, Tramways, Works for the Supply of Gas and Water, and other undertakings of a like character. By L. LIVINGSTON MACASSEY, of the Middle Temple, Barrister-at-Law, and Member of the Institution of Civil Engineers; Author of "Hints on Water Supply." Demy 8vo, 950 pp., price 25s. cloth.

"The volume is a desideratum on a subject which can be only acquired by practical experience, and the order of procedure in Private Bill Legislation and Provisional Orders is followed. The author's suggestions and notes will be found of great value to engineers and others professionally engaged in this class of practice."—*Building News.*

"The author's double experience as an engineer and barrister has eminently qualified him for the task, and enabled him to approach the subject alike from an engineering and legal point of view. The volume will be found a great help both to engineers and lawyers engaged in promoting Private Acts of Parliament and Provisional Orders."—*Local Government Chronicle.*

OGDEN, SMALE AND CO. LIMITED, PRINTERS, GREAT SAFFRON HILL, E.C.

Weale's Rudimentary Series.

LONDON, 1862.
THE PRIZE MEDAL
Was awarded to the Publishers of
"WEALE'S SERIES."

A NEW LIST OF
WEALE'S SERIES
RUDIMENTARY SCIENTIFIC, EDUCATIONAL, AND CLASSICAL.

Comprising nearly Three Hundred and Fifty distinct works in almost every department of Science, Art, and Education, recommended to the notice of Engineers, Architects, Builders, Artisans, and Students generally, as well as to those interested in Workmen's Libraries, Literary and Scientific Institutions, Colleges, Schools, Science Classes, &c., &c.

☞ "WEALE'S SERIES includes Text-Books on almost every branch of Science and Industry, comprising such subjects as Agriculture, Architecture and Building, Civil Engineering, Fine Arts, Mechanics and Mechanical Engineering, Physical and Chemical Science, and many miscellaneous Treatises. The whole are constantly undergoing revision, and new editions, brought up to the latest discoveries in scientific research, are constantly issued. The prices at which they are sold are as low as their excellence is assured."—*American Literary Gazette*.

"Amongst the literature of technical education, WEALE'S SERIES has ever enjoyed a high reputation, and the additions being made by Messrs. CROSBY LOCKWOOD & SON render the series even more complete, and bring the information upon the several subjects down to the present time."—*Mining Journal*.

"It is not too much to say that no books have ever proved more popular with, or more useful to, young engineers and others than the excellent treatises comprised in WEALE'S SERIES."—*Engineer*.

"The excellence of WEALE'S SERIES is now so well appreciated, that it would be wasting our space to enlarge upon their general usefulness and value."—*Builder*.

"WEALE'S SERIES has become a standard as well as an unrivalled collection of treatises in all branches of art and science."—*Public Opinion*.

PHILADELPHIA, 1876.
THE PRIZE MEDAL
Was awarded to the Publishers for
Books: Rudimentary, Scientific,
"WEALE'S SERIES," ETC.

CROSBY LOCKWOOD & SON,
7, STATIONERS' HALL COURT, LUDGATE HILL, LONDON, E.C.

WEALE'S RUDIMENTARY SCIENTIFIC SERIES.

*** The volumes of this Series are freely Illustrated with Woodcuts, or otherwise, where requisite. Throughout the following List it must be understood that the books are bound in limp cloth, unless otherwise stated; *but the volumes marked with a ‡ may also be had strongly bound in cloth boards for 6d. extra.*

N.B.—In ordering from this List it is recommended, as a means of facilitating business and obviating error, to quote the numbers affixed to the volumes, as well as the titles and prices.

CIVIL ENGINEERING, SURVEYING, ETC.

No.
31. *WELLS AND WELL-SINKING.* By JOHN GEO. SWINDELL, A.R.I.B.A., and G. R. BURNELL, C.E. Revised Edition. With a New Appendix on the Qualities of Water. Illustrated. 2s.
35. *THE BLASTING AND QUARRYING OF STONE,* for Building and other Purposes. By Gen. Sir J. BURGOYNE, Bart. 1s. 6d.
43. *TUBULAR, AND OTHER IRON GIRDER BRIDGES,* particularly describing the Britannia and Conway Tubular Bridges. By G. DRYSDALE DEMPSEY, C.E. Fourth Edition. 2s.
44. *FOUNDATIONS AND CONCRETE WORKS,* with Practical Remarks on Footings, Sand, Concrete, Béton, Pile-driving, Caissons, and Cofferdams, &c. By E. DOBSON. Fifth Edition. 1s. 6d.
60. *LAND AND ENGINEERING SURVEYING.* By T. BAKER, C.E. Fourteenth Edition, revised by Professor J. R. YOUNG. 2s.‡
80*. *EMBANKING LANDS FROM THE SEA.* With examples and Particulars of actual Embankments, &c. By J. WIGGINS, F.G.S. 2s.
81. *WATER WORKS,* for the Supply of Cities and Towns. With a Description of the Principal Geological Formations of England as influencing Supplies of Water, &c. By S. HUGHES, C.E. New Edition. 4s.‡
118. *CIVIL ENGINEERING IN NORTH AMERICA,* a Sketch of. By DAVID STEVENSON, F.R.S.E., &c. Plates and Diagrams. 3s.
167. *IRON BRIDGES, GIRDERS, ROOFS, AND OTHER WORKS.* By FRANCIS CAMPIN, C.E. 2s. 6d.‡
197. *ROADS AND STREETS.* By H. LAW, C.E., revised and enlarged by D. K. CLARK, C.E., including pavements of Stone, Wood, Asphalte, &c. 4s. 6d.‡
203. *SANITARY WORK IN THE SMALLER TOWNS AND IN VILLAGES.* By C. SLAGG, A.M.I.C.E. Revised Edition. 3s.‡
212. *GAS-WORKS, THEIR CONSTRUCTION AND ARRANGEMENT;* and the Manufacture and Distribution of Coal Gas. Originally written by SAMUEL HUGHES, C.E. Re-written and enlarged by WILLIAM RICHARDS, C.E. Seventh Edition, with important additions. 5s. 6d.‡
213. *PIONEER ENGINEERING.* A Treatise on the Engineering Operations connected with the Settlement of Waste Lands in New Countries. By EDWARD DOBSON, Assoc. Inst. C.E. 4s. 6d.‡
216. *MATERIALS AND CONSTRUCTION;* A Theoretical and Practical Treatise on the Strains, Designing, and Erection of Works of Construction. By FRANCIS CAMPIN, C.E. Second Edition, revised. 3s.‡
219. *CIVIL ENGINEERING.* By HENRY LAW, M.Inst. C.E. Including HYDRAULIC ENGINEERING by GEO. R. BURNELL, M.Inst. C.E. Seventh Edition, revised, with large additions by D. KINNEAR CLARK, M.Inst. C.E. 6s. 6d., Cloth boards, 7s. 6d.
268. *THE DRAINAGE OF LANDS, TOWNS, & BUILDINGS.* By G. D. DEMPSEY, C.E. Revised, with large Additions on Recent Practice in Drainage Engineering, by D. KINNEAR CLARK, M.I.C.E. Second Edition, Corrected. 4s. 6d.‡ *(Just published.*

☞ *The ‡ indicates that these vols. may be had strongly bound at 6d. extra.*

LONDON: CROSBY LOCKWOOD AND SON.

MECHANICAL ENGINEERING, ETC.

33. *CRANES,* the Construction of, and other Machinery for Raising Heavy Bodies. By JOSEPH GLYNN, F.R.S. Illustrated. 1s. 6d.
34. *THE STEAM ENGINE.* By Dr. LARDNER. Illustrated. 1s. 6d.
59. *STEAM BOILERS:* their Construction and Management. By R. ARMSTRONG, C.E. Illustrated. 1s. 6d.
82. *THE POWER OF WATER,* as applied to drive Flour Mills, and to give motion to Turbines, &c. By JOSEPH GLYNN, F.R.S. 2s.‡
98. *PRACTICAL MECHANISM,* the Elements of; and Machine Tools. By T. BAKER, C.E. With Additions by J. NASMYTH, C.E. 2s. 6d.‡
139. *THE STEAM ENGINE,* a Treatise on the Mathematical Theory of, with Rules and Examples for Practical Men. By T. BAKER, C.E. 1s. 6d.
164. *MODERN WORKSHOP PRACTICE,* as applied to Steam Engines, Bridges, Ship-building, Cranes, &c. By J. G. WINTON. Fourth Edition, much enlarged and carefully revised. 3s. 6d.‡ [*Just published.*
165. *IRON AND HEAT,* exhibiting the Principles concerned in the Construction of Iron Beams, Pillars, and Girders. By J. ARMOUR. 2s. 6d.‡
166. *POWER IN MOTION:* Horse-Power, Toothed-Wheel Gearing, Long and Short Driving Bands, and Angular Forces. By J. ARMOUR, 2s.‡
171. *THE WORKMAN'S MANUAL OF ENGINEERING DRAWING.* By J. MAXTON. 6th Edn. With 7 Plates and 350 Cuts. 3s. 6d.‡
190. *STEAM AND THE STEAM ENGINE,* Stationary and Portable. Being an Extension of the Elementary Treatise on the Steam Engine of Mr. JOHN SEWELL. By D. K. CLARK, M.I.C.E. 3s. 6d.‡
200. *FUEL,* its Combustion and Economy. By C. W. WILLIAMS. With Recent Practice in the Combustion and Economy of Fuel—Coal, Coke, Wood, Peat, Petroleum, &c.—by D. K. CLARK, M.I.C.E. 3s. 6d.‡
202. *LOCOMOTIVE ENGINES.* By G. D. DEMPSEY, C.E.; with large additions by D. KINNEAR CLARK, M.I.C.E. 3s.‡
211. *THE BOILERMAKER'S ASSISTANT* in Drawing, Templating, and Calculating Boiler and Tank Work. By JOHN COURTNEY. Practical Boiler Maker. Edited by D. K. CLARK, C.E. 100 Illustrations. 2s,
217. *SEWING MACHINERY:* Its Construction, History, &c., with full Technical Directions for Adjusting, &c. By J. W. URQUHART, C.E. 2s.‡
223. *MECHANICAL ENGINEERING.* Comprising Metallurgy, Moulding, Casting, Forging, Tools, Workshop Machinery, Manufacture of the Steam Engine, &c. By FRANCIS CAMPIN, C.E. Second Edition. 2s. 6d.‡
236. *DETAILS OF MACHINERY.* Comprising Instructions for the Execution of various Works in Iron. By FRANCIS CAMPIN, C.E. 3s.‡
237. *THE SMITHY AND FORGE;* including the Farrier's Art and Coach Smithing. By W. J. E. CRANE. Illustrated. 2s. 6d.‡
238. *THE SHEET-METAL WORKER'S GUIDE;* a Practical Handbook for Tinsmiths, Coppersmiths, Zincworkers, &c. With 94 Diagrams and Working Patterns. By W. J. E. CRANE. Second Edition, revised. 1s. 6d.
251. *STEAM AND MACHINERY MANAGEMENT:* with Hints on Construction and Selection. By M. POWIS BALE, M.I.M E. 2s. 6d.‡
254. *THE BOILERMAKER'S READY-RECKONER.* By J. COURTNEY. Edited by D. K. CLARK, C.E. 4s., limp; 5s., half-bound.
255. *LOCOMOTIVE ENGINE-DRIVING.* A Practical Manual for Engineers in charge of Locomotive Engines. By MICHAEL REYNOLDS, M.S.E Eighth Edition. 3s. 6d., limp; 4s. 6d. cloth boards.
256. *STATIONARY ENGINE-DRIVING.* A Practical Manual Engineers in charge of Stationary Engines. By MICHAEL REYNOLDS, M.S.E. Third Edition. 3s. 6d. limp; 4s. 6d. cloth boards.
260. *IRON BRIDGES OF MODERATE SPAN:* their Construction and Erection. By HAMILTON W. PENDRED, C.E. 2s.

☞ *The ‡ indicates that these vols. may be had strongly bound at 6d. extra.*

7, STATIONERS' HALL COURT, LUDGATE HILL, E.C.

MINING, METALLURGY, ETC.

4. *MINERALOGY*, Rudiments of; a concise View of the General Properties of Minerals. By A. RAMSAY, F.G.S., F.R.G.S., &c. Third Edition, revised and enlarged. Illustrated. 3s. 6d.‡
117. *SUBTERRANEOUS SURVEYING*, with and without the Magnetic Needle. By T. FENWICK and T. BAKER, C.E. Illustrated. 2s. 6d.‡
133. *METALLURGY OF COPPER*. By R. H. LAMBORN. 2s. 6d.‡
135. *ELECTRO-METALLURGY*; Practically Treated. By ALEXANDER WATT. Ninth Edition, enlarged and revised, with additional Illustrations, and including the most recent Processes. 3s. 6d.‡
172. *MINING TOOLS*, Manual of. For the Use of Mine Managers, Agents, Students, &c. By WILLIAM MORGANS. 2s. 6d.
172*. *MINING TOOLS, ATLAS* of Engravings to Illustrate the above, containing 235 Illustrations, drawn to Scale. 4to. 4s. 6d.
176. *METALLURGY OF IRON*. Containing History of Iron Manufacture, Methods of Assay, and Analyses of Iron Ores, Processes of Manufacture of Iron and Steel, &c. By H. BAUERMAN, F.G.S. Sixth Edition, revised and enlarged. 5s.‡ [*Just published*.
180. *COAL AND COAL MINING*. By SIR WARINGTON W. SMYTH, M.A., F.R.S. Seventh Edition, revised. 3s. 6d.‡ [*Just published*.
195. *THE MINERAL SURVEYOR AND VALUER'S COMPLETE GUIDE*. By W. LINTERN, Mining Engineer. Third Edition, with an Appendix on Magnetic and Angular Surveying. With Four Plates. 3s. 6d.‡ [*Just published*.
214. *SLATE AND SLATE QUARRYING*, Scientific, Practical, and Commercial. By D. C. DAVIES, F.G.S., Mining Engineer, &c. 3s.‡
264. *A FIRST BOOK OF MINING AND QUARRYING*, with the Sciences connected therewith, for Primary Schools and Self Instruction. By J. H. COLLINS, F.G.S. Second Edition, with additions. 1s. 6d.

ARCHITECTURE, BUILDING, ETC.

16. *ARCHITECTURE—ORDERS*—The Orders and their Æsthetic Principles. By W. H. LEEDS. Illustrated. 1s. 6d.
17. *ARCHITECTURE—STYLES*—The History and Description of the Styles of Architecture of Various Countries, from the Earliest to the Present Period. By T. TALBOT BURY, F.R.I.B.A., &c. Illustrated. 2s.
*** ORDERS AND STYLES OF ARCHITECTURE, *in One Vol*., 3s. 6d.
18. *ARCHITECTURE—DESIGN*—The Principles of Design in Architecture, as deducible from Nature and exemplified in the Works of the Greek and Gothic Architects. By E. L. GARBETT, Architect. Illustrated. 2s.6d.
*** The three preceding Works, in One handsome Vol., half bound, entitled "MODERN ARCHITECTURE," *price* 6s.
22. *THE ART OF BUILDING*, Rudiments of. General Principles of Construction, Materials used in Building, Strength and Use of Materials, Working Drawings, Specifications, and Estimates. By E. DOBSON, 2s.‡
25. *MASONRY AND STONECUTTING*: Rudimentary Treatise on the Principles of Masonic Projection and their application to Construction. By EDWARD DOBSON, M.R.I.B.A., &c. 2s. 6d.‡
42. *COTTAGE BUILDING*. By C. BRUCE ALLEN, Architect. Tenth Edition, revised and enlarged. With a Chapter on Economic Cottages for Allotments, by EDWARD E. ALLEN, C.E. 2s.
45. *LIMES, CEMENTS, MORTARS, CONCRETES, MASTICS*, PLASTERING, &c. By G. R. BURNELL, C.E. Thirteenth Edition. 1s. 6d.
57. *WARMING AND VENTILATION*. An Exposition of the General Principles as applied to Domestic and Public Buildings, Mines, Lighthouses, Ships, &c. By C. TOMLINSON, F.R.S., &c. Illustrated. 3s.

☞ The ‡ *indicates that these vols. may be had strongly bound at* 6d. *extra*.

LONDON : CROSBY LOCKWOOD AND SON,

Architecture, Building, etc., *continued.*

111. *ARCHES, PIERS, BUTTRESSES, &c.:* Experimental Essays on the Principles of Construction. By W. BLAND. Illustrated. 1s. 6d.
116. *THE ACOUSTICS OF PUBLIC BUILDINGS;* or, The Principles of the Science of Sound applied to the purposes of the Architect and Builder. By T. ROGER SMITH, M.R.I.B.A., Architect. Illustrated. 1s. 6d.
127. *ARCHITECTURAL MODELLING IN PAPER,* the Art of. By T. A. RICHARDSON, Architect. Illustrated. 1s. 6d.
128. *VITRUVIUS — THE ARCHITECTURE OF MARCUS VITRUVIUS POLLO.* In Ten Books. Translated from the Latin by JOSEPH GWILT, F.S.A., F.R.A.S. With 23 Plates. 5s.
130. *GRECIAN ARCHITECTURE,* An Inquiry into the Principles of Beauty in; with an Historical View of the Rise and Progress of the Art in Greece. By the EARL OF ABERDEEN. 1s.
⁎⁎* The two preceding Works in One handsome Vol., half bound, entitled "ANCIENT ARCHITECTURE," price 6s.*
132. *THE ERECTION OF DWELLING-HOUSES.* Illustrated by a Perspective View, Plans, Elevations, and Sections of a pair of Semi-detached Villas, with the Specification, Quantities, and Estimates, &c. By S. H. BROOKS. New Edition, with Plates. 2s. 6d.‡
156. *QUANTITIES & MEASUREMENTS* in Bricklayers', Masons', Plasterers', Plumbers', Painters', Paperhangers', Gilders', Smiths', Carpenters' and Joiners' Work. By A. C. BEATON, Surveyor. New Edition. 1s. 6d.
175. *LOCKWOOD'S BUILDER'S PRICE BOOK FOR* 1890. A Comprehensive Handbook of the Latest Prices and Data for Builders, Architects, Engineers, and Contractors. Re-constructed, Re-written, and greatly Enlarged. By FRANCIS T. W. MILLER, A.R.I.B.A. 640 pages. 3s. 6d.‡ [*Just published.*
182. *CARPENTRY AND JOINERY* — THE ELEMENTARY PRINCIPLES OF CARPENTRY. Chiefly composed from the Standard Work of THOMAS TREDGOLD, C.E. With a TREATISE ON JOINERY by E. WYNDHAM TARN, M.A. Fourth Edition, Revised. 3s. 6d.‡
182*. *CARPENTRY AND JOINERY. ATLAS* of 35 Plates to accompany the above. With Descriptive Letterpress. 4to. 6s.
185. *THE COMPLETE MEASURER;* the Measurement of Boards, Glass. &c.; Unequal-sided, Square-sided, Octagonal-sided, Round Timber and Stone, and Standing Timber, &c. By RICHARD HORTON. Fifth Edition. 4s.; strongly bound in leather, 5s.
187. *HINTS TO YOUNG ARCHITECTS.* By G. WIGHTWICK. New Edition. By G. H. GUILLAUME. Illustrated. 3s. 6d.‡
188. *HOUSE PAINTING, GRAINING, MARBLING, AND SIGN WRITING:* with a Course of Elementary Drawing for House-Painters, Sign-Writers, &c., and a Collection of Useful Receipts. By ELLIS A. DAVIDSON. Fifth Edition. With Coloured Plates. 5s. cloth limp; 6s. cloth boards.
189. *THE RUDIMENTS OF PRACTICAL BRICKLAYING.* In Six Sections: General Principles; Arch Drawing, Cutting, and Setting; Pointing; Paving. Tiling, Materials; Slating and Plastering; Practical Geometry, Mensuration, &c. By ADAM HAMMOND. Seventh Edition. 1s. 6d.
191. *PLUMBING.* A Text-Book to the Practice of the Art or Craft of the Plumber. With Chapters upon House Drainage and Ventilation. Fifth Edition. With 380 Illustrations. By W. P. BUCHAN. 3s. 6d.‡
192. *THE TIMBER IMPORTER'S, TIMBER MERCHANT'S,* and BUILDER'S STANDARD GUIDE. By R. E. GRANDY. 2s.
206. *A BOOK ON BUILDING, Civil and Ecclesiastical,* including CHURCH RESTORATION. With the Theory of Domes and the Great Pyramid, &c. By Sir EDMUND BECKETT, Bart., LL.D., Q.C., F.R.A.S. 4s. 6d.‡
226. *THE JOINTS MADE AND USED BY BUILDERS* in the Construction of various kinds of Engineering and Architectural Works. By WYVILL J. CHRISTY, Architect. With upwards of 160 Engravings on Wood. 3s.‡

☞ *The ‡ indicates that these vols. may be had strongly bound at 6d. extra.*

7, STATIONERS' HALL COURT, LUDGATE HILL, E.C.

Architecture, Building, etc., *continued*.

228. *THE CONSTRUCTION OF ROOFS OF WOOD AND IRON.* By E. WYNDHAM TARN, M.A., Architect. Second Edition, revised. 1s. 6d.

229. *ELEMENTARY DECORATION:* as applied to the Interior and Exterior Decoration of Dwelling-Houses, &c. By J. W. FACEY. 2s.

257. *PRACTICAL HOUSE DECORATION.* A Guide to the Art of Ornamental Painting. By JAMES W. FACEY. 2s. 6d.

** *The two preceding Works, in One handsome Vol., half-bound, entitled* "HOUSE DECORATION, ELEMENTARY AND PRACTICAL," *price* 5s.

230. *HANDRAILING.* Showing New and Simple Methods for finding the Pitch of the Plank. Drawing the Moulds, Bevelling, Jointing-up, and Squaring the Wreath. By GEORGE COLLINGS. Plates and Diagrams. 1s. 6d.

247. *BUILDING ESTATES:* a Rudimentary Treatise on the Development, Sale, Purchase, and General Management of Building Land. By FOWLER MAITLAND, Surveyor. Second Edition, revised. 2s.

248. *PORTLAND CEMENT FOR USERS.* By HENRY FAIJA, Assoc. M. Inst. C.E. Second Edition, corrected. Illustrated. 2s.

252. *BRICKWORK:* a Practical Treatise, embodying the General and Higher Principles of Bricklaying, Cutting and Setting, &c. By F. WALKER. Second Edition, Revised and Enlarged. 1s. 6d.

23. *THE PRACTICAL BRICK AND TILE BOOK.* Comprising:
189. BRICK AND TILE MAKING, by E. DOBSON, A.I.C.E.; PRACTICAL BRICKLAY-
252. ING, by A. HAMMOND; BRICKWORK, by F. WALKER. 550 pp. with 270 Illustrations. 6s. Strongly half-bound.

253. *THE TIMBER MERCHANT'S, SAW-MILLER'S, AND IMPORTER'S FREIGHT-BOOK AND ASSISTANT.* By WM. RICHARDSON. With a Chapter on Speeds of Saw-Mill Machinery, &c. By M. POWIS BALE, A.M.Inst.C.E. 3s.‡

258. *CIRCULAR WORK IN CARPENTRY AND JOINERY.* A Practical Treatise on Circular Work of Single and Double Curvature. By GEORGE COLLINGS, Author of "A Treatise on Handrailing." 2s. 6d.

259. *GAS FITTING:* A Practical Handbook treating of every Description of Gas Laying and Fitting. By JOHN BLACK. With 122 Illustrations. 2s. 6d.‡

261. *SHORING AND ITS APPLICATION:* A Handbook for the Use of Students. By GEORGE H. BLAGROVE. 1s. 6d. [*Just published*.

265. *THE ART OF PRACTICAL BRICK CUTTING & SETTING.* By ADAM HAMMOND. With 90 Engravings. 1s. 6d. [*Just published*.

267. *THE SCIENCE OF BUILDING:* An Elementary Treatise on the Principles of Construction. Adapted to the Requirements of Architectural Students. By E. WYNDHAM TARN, M.A. Lond. Third Edition, Revised and Enlarged. With 59 Wood Engravings. 3s. 6d.‡ [*Just published*.

SHIPBUILDING, NAVIGATION, MARINE ENGINEERING, ETC.

51. *NAVAL ARCHITECTURE.* An Exposition of the Elementary Principles of the Science, and their Practical Application to Naval Construction. By J. PEAKE. Fifth Edition, with Plates and Diagrams. 3s. 6d.‡

53*. *SHIPS FOR OCEAN & RIVER SERVICE,* Elementary and Practical Principles of the Construction of. By H. A. SOMMERFELDT. 1s. 6d.

53**. *AN ATLAS OF ENGRAVINGS* to Illustrate the above. Twelve large folding plates. Royal 4to, cloth. 7s. 6d.

54. *MASTING, MAST-MAKING, AND RIGGING OF SHIPS,* Also Tables of Spars, Rigging, Blocks; Chain, Wire, and Hemp Ropes, &c., relative to every class of vessels. By ROBERT KIPPING, N.A. 2s.

54*. *IRON SHIP-BUILDING.* With Practical Examples and Details. By JOHN GRANTHAM, C.E. 5th Edition. 4s.

☞ *The* ‡ *indicates that these vols. may be had strongly bound at 6d. extra.*

LONDON: CROSBY LOCKWOOD AND SON,

WEALE'S RUDIMENTARY SERIES.

Shipbuilding, Navigation, Marine Engineering, etc., *cont.*

55. *THE SAILOR'S SEA BOOK:* a Rudimentary Treatise on Navigation. By JAMES GREENWOOD, B.A. With numerous Woodcuts and Coloured Plates. New and enlarged edition. By W. H. ROSSER. 2s. 6d.‡
80. *MARINE ENGINES AND STEAM VESSELS.* By ROBERT MURRAY, C.E. Eighth Edition, thoroughly Revised, with Additions by the Author and by GEORGE CARLISLE, C.E., Senior Surveyor to the Board of Trade, Liverpool. 4s. 6d. limp; 5s. cloth boards.
83*bis*. *THE FORMS OF SHIPS AND BOATS.* By W. BLAND. Seventh Edition, Revised, with numerous Illustrations and Models. 1s. 6d.
99. *NAVIGATION AND NAUTICAL ASTRONOMY*, in Theory and Practice. By Prof. J. R. YOUNG. New Edition. 2s. 6d.
106. *SHIPS' ANCHORS*, a Treatise on. By G. COTSELL, N.A. 1s. 6d.
149. *SAILS AND SAIL-MAKING.* With Draughting, and the Centre of Effort of the Sails; Weights and Sizes of Ropes; Masting, Rigging, and Sails of Steam Vessels, &c. 12th Edition. By R. KIPPING. N.A. 2s. 6d.‡
155. *ENGINEER'S GUIDE TO THE ROYAL & MERCANTILE NAVIES.* By a PRACTICAL ENGINEER. Revised by D. F. M'CARTHY. 3s.
55 & 204. *PRACTICAL NAVIGATION.* Consisting of The Sailor's Sea-Book. By JAMES GREENWOOD and W. H. ROSSER. Together with the requisite Mathematical and Nautical Tables for the Working of the Problems. By H. LAW, C.E., and Prof. J. R. YOUNG. 7s. Half-bound.

AGRICULTURE, GARDENING, ETC.

61*. *A COMPLETE READY RECKONER FOR THE ADMEASUREMENT OF LAND*, &c. By A. ARMAN. Third Edition, revised and extended by C. NORRIS, Surveyor, Valuer, &c. 2s.
131. *MILLER'S, CORN MERCHANT'S, AND FARMER'S READY RECKONER.* Second Edition, with a Price List of Modern Flour-Mill Machinery, by W. S. HUTTON, C.E. 2s.
140. *SOILS, MANURES, AND CROPS.* (Vol. 1. OUTLINES OF MODERN FARMING.) By R. SCOTT BURN. Woodcuts. 2s.
141. *FARMING & FARMING ECONOMY*, Notes, Historical and Practical, on. (Vol. 2. OUTLINES OF MODERN FARMING.) By R. SCOTT BURN. 3s.
142. *STOCK; CATTLE, SHEEP, AND HORSES.* (Vol. 3. OUTLINES OF MODERN FARMING.) By R. SCOTT BURN. Woodcuts. 2s. 6d.
145. *DAIRY, PIGS, AND POULTRY*, Management of the. By R. SCOTT BURN. (Vol. 4. OUTLINES OF MODERN FARMING.) 2s.
146. *UTILIZATION OF SEWAGE, IRRIGATION, AND RECLAMATION OF WASTE LAND.* (Vol. 5. OUTLINES OF MODERN FARMING.) By R. SCOTT BURN. Woodcuts. 2s. 6d.
⁎ Nos. 140-1-2-5-6, *in One Vol., handsomely half-bound, entitled* "OUTLINES OF MODERN FARMING." By ROBERT SCOTT BURN. *Price* 12s.
177. *FRUIT TREES*, The Scientific and Profitable Culture of. From the French of DU BREUIL. Revised by GEO. GLENNY. 187 Woodcuts. 3s. 6d.‡
198. *SHEEP; THE HISTORY, STRUCTURE, ECONOMY, AND DISEASES OF.* By W. C. SPOONER, M.R.V.C., &c. Fifth Edition, enlarged, including Specimens of New and Improved Breeds. 3s. 6d.‡
201. *KITCHEN GARDENING MADE EASY.* By GEORGE M. F. GLENNY. Illustrated. 1s. 6d.‡
207. *OUTLINES OF FARM MANAGEMENT, and the Organization of Farm Labour.* By R. SCOTT BURN. 2s. 6d.‡
208. *OUTLINES OF LANDED ESTATES MANAGEMENT.* By R. SCOTT BURN. 2s. 6d.‡
⁎ Nos. 207 & 208 *in One Vol., handsomely half-bound, entitled* "OUTLINES OF LANDED ESTATES AND FARM MANAGEMENT." By R. SCOTT BURN. *Price* 6s.

☞ *The ‡ indicates that these vols. may be had strongly bound at 6d. extra.*

7, STATIONERS' HALL COURT, LUDGATE HILL, E.C.

Agriculture, Gardening, etc., *continued.*

209. *THE TREE PLANTER AND PLANT PROPAGATOR.*
A Practical Manual on the Propagation of Forest Trees, Fruit Trees, Flowering Shrubs, Flowering Plants, &c. By SAMUEL WOOD. 2s.‡

210. *THE TREE PRUNER.* A Practical Manual on the Pruning of Fruit Trees, including also their Training and Renovation; also the Pruning of Shrubs, Climbers, and Flowering Plants. By SAMUEL WOOD. 2s.‡

*** *Nos. 209 & 210 in One Vol., handsomely half-bound, entitled* "THE TREE PLANTER, PROPAGATOR, AND PRUNER." By SAMUEL WOOD. *Price* 5s.

218. *THE HAY AND STRAW MEASURER:* Being New Tables for the Use of Auctioneers, Valuers, Farmers, Hay and Straw Dealers, &c. By JOHN STEELE. Fourth Edition. 2s.

222. *SUBURBAN FARMING.* The Laying-out and Cultivation of Farms, adapted to the Produce of Milk, Butter, and Cheese, Eggs, Poultry, and Pigs. By Prof. JOHN DONALDSON and R. SCOTT BURN. 3s. 6d.‡

231. *THE ART OF GRAFTING AND BUDDING.* By CHARLES BALTET. With Illustrations. 2s. 6d.‡

232. *COTTAGE GARDENING;* or, Flowers, Fruits, and Vegetables for Small Gardens. By E. HOBDAY. 1s. 6d.

233. *GARDEN RECEIPTS.* Edited by CHARLES W. QUIN. 1s. 6d.

234. *MARKET AND KITCHEN GARDENING.* By C. W. SHAW, late Editor of "Gardening Illustrated." 3s.‡ [*Just published.*]

239. *DRAINING AND EMBANKING.* A Practical Treatise, embodying the most recent experience in the Application of Improved Methods. By JOHN SCOTT, late Professor of Agriculture and Rural Economy at the Royal Agricultural College, Cirencester. With 68 Illustrations. 1s. 6d.

240. *IRRIGATION AND WATER SUPPLY.* A Treatise on Water Meadows, Sewage Irrigation, and Warping; the Construction of Wells, Ponds, and Reservoirs, &c. By Prof. JOHN SCOTT. With 34 Illus. 1s. 6d.

241. *FARM ROADS, FENCES, AND GATES.* A Practical Treatise on the Roads, Tramways, and Waterways of the Farm; the Principles of Enclosures; and the different kinds of Fences, Gates, and Stiles. By Professor JOHN SCOTT. With 75 Illustrations. 1s. 6d.

242. *FARM BUILDINGS.* A Practical Treatise on the Buildings necessary for various kinds of Farms, their Arrangement and Construction, with Plans and Estimates. By Prof. JOHN SCOTT. With 105 Illus. 2s.

243. *BARN IMPLEMENTS AND MACHINES.* A Practical Treatise on the Application of Power to the Operations of Agriculture; and on various Machines used in the Threshing-barn, in the Stock-yard, and in the Dairy, &c. By Prof. J. SCOTT. With 123 Illustrations. 2s.

244. *FIELD IMPLEMENTS AND MACHINES.* A Practical Treatise on the Varieties now in use, with Principles and Details of Construction, their Points of Excellence, and Management. By Professor JOHN SCOTT. With 138 Illustrations. 2s.

245. *AGRICULTURAL SURVEYING.* A Practical Treatise on Land Surveying, Levelling, and Setting-out; and on Measuring and Estimating Quantities, Weights, and Values of Materials, Produce, Stock, &c. By Prof. JOHN SCOTT. With 62 Illustrations. 1s. 6d.

*** *Nos. 239 to 245 in One Vol., handsomely half-bound, entitled* "THE COMPLETE TEXT-BOOK OF FARM ENGINEERING." By Professor JOHN SCOTT. *Price* 12s.

250. *MEAT PRODUCTION.* A Manual for Producers, Distributors, &c. By JOHN EWART. 2s. 6d.‡

266. *BOOK-KEEPING FOR FARMERS & ESTATE OWNERS.* By J. M. WOODMAN, Chartered Accountant. 2s. 6d. cloth limp; 3s. 6d. cloth boards. [*Just published.*]

The ‡ *indicates that these vols. may be had strongly bound at* 6d. *extra.*

MATHEMATICS, ARITHMETIC, ETC.

32. *MATHEMATICAL INSTRUMENTS*, a Treatise on; Their Construction, Adjustment, Testing, and Use concisely Explained. By J. F. HEATHER, M.A. Fourteenth Edition, revised, with additions, by A. T. WALMISLEY, M.I.C.E., Fellow of the Surveyors' Institution. Original Edition, in 1 vol., Illustrated. 2s.‡ [*Just published.*
⁎⁎ *In ordering the above, be careful to say, " Original Edition " (No. 32), to distinguish it from the Enlarged Edition in 3 vols. (Nos. 168-9-70.)*

76. *DESCRIPTIVE GEOMETRY*, an Elementary Treatise on; with a Theory of Shadows and of Perspective, extracted from the French of G. MONGE. To which is added, a description of the Principles and Practice of Isometrical Projection. By J. F. HEATHER, M.A. With 14 Plates. 2s.

178. *PRACTICAL PLANE GEOMETRY:* giving the Simplest Modes of Constructing Figures contained in one Plane and Geometrical Construction of the Ground. By J. F. HEATHER, M.A. With 215 Woodcuts. 2s.

83. *COMMERCIAL BOOK-KEEPING.* With Commercial Phrases and Forms in English, French, Italian, and German. By JAMES HADDON, M.A., Arithmetical Master of King's College School, London. 1s. 6d.

84. *ARITHMETIC*, a Rudimentary Treatise on: with full Explanations of its Theoretical Principles, and numerous Examples for Practice. By Professor J. R. YOUNG. Eleventh Edition. 1s. 6d.

84*. A KEY to the above, containing Solutions in full to the Exercises, together with Comments, Explanations, and Improved Processes, for the Use of Teachers and Unassisted Learners. By J. R. YOUNG. 1s. 6d.

85. *EQUATIONAL ARITHMETIC,* applied to Questions of Interest, Annuities, Life Assurance, and General Commerce; with various Tables by which all Calculations may be greatly facilitated. By W. HIPSLEY. 2s.

86. *ALGEBRA*, the Elements of. By JAMES HADDON, M.A. With Appendix, containing miscellaneous Investigations, and a Collection of Problems in various parts of Algebra. 2s.

86*. A KEY AND COMPANION to the above Book, forming an extensive repository of Solved Examples and Problems in Illustration of the various Expedients necessary in Algebraical Operations. By J. R. YOUNG. 1s. 6d.

88. *EUCLID*, THE ELEMENTS OF: with many additional Propositions
89. and Explanatory Notes: to which is prefixed, an Introductory Essay on Logic. By HENRY LAW, C.E. 2s. 6d.‡

⁎⁎ *Sold also separately, viz. :—*

88. EUCLID, The First Three Books. By HENRY LAW, C.E. 1s. 6d.
89. EUCLID, Books 4, 5, 6, 11, 12. By HENRY LAW, C.E. 1s. 6d.

90. *ANALYTICAL GEOMETRY AND CONIC SECTIONS*, By JAMES HANN. A New Edition, by Professor J. R. YOUNG. 2s.‡

91. *PLANE TRIGONOMETRY,* the Elements of. By JAMES HANN, formerly Mathematical Master of King's College, London. 1s. 6d.

92. *SPHERICAL TRIGONOMETRY,* the Elements of. By JAMES HANN. Revised by CHARLES H. DOWLING, C.E. 1s.
⁎⁎ *Or with " The Elements of Plane Trigonometry," in One Volume, 2s. 6d.*

93. *MENSURATION AND MEASURING.* With the Mensuration and Levelling of Land for the Purposes of Modern Engineering. By T. BAKER, C.E. New Edition by E. NUGENT, C.E. Illustrated. 1s. 6d.

101. *DIFFERENTIAL CALCULUS,* Elements of the. By W. S. B. WOOLHOUSE, F.R.A.S., &c. 1s. 6d.

102. *INTEGRAL CALCULUS,* Rudimentary Treatise on the. By HOMERSHAM COX, B.A. Illustrated. 1s.

136. *ARITHMETIC,* Rudimentary, for the Use of Schools and Self-Instruction. By JAMES HADDON, M.A. Revised by A. ARMAN. 1s. 6d.

137. A KEY TO HADDON'S RUDIMENTARY ARITHMETIC. By A. ARMAN. 1s. 6d.

☞ *The ‡ indicates that these vols. may be had strongly bound at 6d. extra.*

7, STATIONERS' HALL COURT, LUDGATE HILL. E.C.

Mathematics, Arithmetic, etc., *continued.*

168. *DRAWING AND MEASURING INSTRUMENTS.* Including—I. Instruments employed in Geometrical and Mechanical Drawing, and in the Construction, Copying, and Measurement of Maps and Plans. II. Instruments used for the purposes of Accurate Measurement, and for Arithmetical Computations. By J. F. HEATHER, M.A. Illustrated. 1s. 6d

169. *OPTICAL INSTRUMENTS.* Including (more especially) Telescopes, Microscopes, and Apparatus for producing copies of Maps and Plans by Photography. By J. F. HEATHER, M.A. Illustrated. 1s. 6d.

170. *SURVEYING AND ASTRONOMICAL INSTRUMENTS.* Including—I. Instruments Used for Determining the Geometrical Features of a portion of Ground. II. Instruments Employed in Astronomical Observations. By J. F. HEATHER, M.A. Illustrated. 1s. 6d.

*** *The above three volumes form an enlargement of the Author's original work "Mathematical Instruments." (See No. 32 in the Series.)*

168.⎫ *MATHEMATICAL INSTRUMENTS.* By J. F. HEATHER,
169. ⎬ M.A. Enlarged Edition, for the most part entirely re-written. The 3 Parts as
170.⎭ above, in One thick Volume. With numerous Illustrations. 4s. 6d.‡

158. *THE SLIDE RULE, AND HOW TO USE IT;* containing full, easy, and simple Instructions to perform all Business Calculations with unexampled rapidity and accuracy. By CHARLES HOARE, C.E. Fifth Edition. With a Slide Rule in tuck of cover. 2s. 6d.‡

196. *THEORY OF COMPOUND INTEREST AND ANNUITIES;* with Tables of Logarithms for the more Difficult Computations of Interest, Discount, Annuities, &c. By FÉDOR THOMAN. 4s.‡

199. *THE COMPENDIOUS CALCULATOR;* or, Easy and Concise Methods of Performing the various Arithmetical Operations required in Commercial and Business Transactions; together with Useful Tables. By D. O'GORMAN. Twenty-seventh Edition, carefully revised by C. NORRIS. 2s. 6d., cloth limp; 3s. 6d., strongly half-bound in leather.

204. *MATHEMATICAL TABLES,* for Trigonometrical, Astronomical, and Nautical Calculations; to which is prefixed a Treatise on Logarithms. By HENRY LAW, C.E. Together with a Series of Tables for Navigation and Nautical Astronomy. By Prof. J. R. YOUNG. New Edition. 4s.

204*. *LOGARITHMS.* With Mathematical Tables for Trigonometrical, Astronomical, and Nautical Calculations. By HENRY LAW, M.Inst.C.E. New and Revised Edition. (Forming part of the above Work). 3s.

221. *MEASURES, WEIGHTS, AND MONEYS OF ALL NATIONS,* and an Analysis of the Christian, Hebrew, and Mahometan Calendars. By W. S. B. WOOLHOUSE, F.R.A.S., F.S.S. Sixth Edition. 2s.‡

227. *MATHEMATICS AS APPLIED TO THE CONSTRUCTIVE ARTS.* Illustrating the various processes of Mathematical Investigation, by means of Arithmetical and Simple Algebraical Equations and Practical Examples. By FRANCIS CAMPIN, C.E. Second Edition. 3s.‡

PHYSICAL SCIENCE, NATURAL PHILOSOPHY, ETC.

1. *CHEMISTRY.* By Professor GEORGE FOWNES, F.R.S. With an Appendix on the Application of Chemistry to Agriculture. 1s.
2. *NATURAL PHILOSOPHY,* Introduction to the Study of. By C. TOMLINSON. Woodcuts. 1s. 6d.
6. *MECHANICS,* Rudimentary Treatise on. By CHARLES TOMLINSON. Illustrated. 1s. 6d.
7. *ELECTRICITY;* showing the General Principles of Electrical Science, and the purposes to which it has been applied. By Sir W. SNOW HARRIS, F.R.S., &c. With Additions by R. SABINE, C.E., F.S.A. 1s. 6d.
7*. *GALVANISM.* By Sir W. SNOW HARRIS. New Edition by ROBERT SABINE, C.E., F.S.A. 1s. 6d.
8. *MAGNETISM;* being a concise Exposition of the General Principles of Magnetical Science. By Sir W. SNOW HARRIS. New Edition, revised by H. M. NOAD, Ph.D. With 165 Woodcuts. 3s. 6d.‡

☞ *The ‡ indicates that these vols. may be had strongly bound at 6d. extra.*

LONDON: CROSBY LOCKWOOD AND SON,

WEALE'S RUDIMENTARY SERIES. 11

Physical Science, Natural Philosophy, etc., *continued.*
11. *THE ELECTRIC TELEGRAPH;* its History and Progress; with Descriptions of some of the Apparatus. By R. SABINE, C.E., F.S.A. 3s.
12. *PNEUMATICS,* including Acoustics and the Phenomena of Wind Currents, for the Use of Beginners By CHARLES TOMLINSON, F.R.S. Fourth Edition, enlarged. Illustrated. 1s. 6d. [*Just published.*
72. *MANUAL OF THE MOLLUSCA;* a Treatise on Recent and Fossil Shells. By Dr. S. P. WOODWARD, A.L.S. Fourth Edition. With Appendix by RALPH TATE, A.L.S., F.G.S. With numerous Plates and 300 Woodcuts. 6s. 6d. Cloth boards, 7s. 6d.
96. *ASTRONOMY.* By the late Rev. ROBERT MAIN, M.A. Third Edition, by WILLIAM THYNNE LYNN, B.A., F.R.A.S. 2s.
97. *STATICS AND DYNAMICS,* the Principles and Practice of; embracing also a clear development of Hydrostatics, Hydrodynamics, and Central Forces. By T. BAKER, C.E. Fourth Edition. 1s. 6d.
138. *TELEGRAPH,* Handbook of the; a Guide to Candidates for Employment in the Telegraph Service. By R. BOND. 3s.‡
173. *PHYSICAL GEOLOGY,* partly based on Major-General PORTLOCK'S "Rudiments of Geology." By RALPH TATE, A.L.S., &c. Woodcuts. 2s.
174. *HISTORICAL GEOLOGY,* partly based on Major-General PORTLOCK'S "Rudiments." By RALPH TATE, A.L.S., &c. Woodcuts. 2s. 6d.
173 *RUDIMENTARY TREATISE ON GEOLOGY,* Physical and
& Historical. Partly based on Major-General PORTLOCK'S "Rudiments of
174. Geology." By RALPH TATE, A.L.S., F.G.S., &c. In One Volume. 4s. 6d.‡
183 *ANIMAL PHYSICS,* Handbook of. By Dr. LARDNER, D.C.L.,
& formerly Professor of Natural Philosophy and Astronomy in University
184. College, Lond. With 520 Illustrations. In One Vol. 7s. 6d., cloth boards.
*** *Sold also in Two Parts, as follows :—*
183. ANIMAL PHYSICS. By Dr. LARDNER. Part I., Chapters I.—VII. 4s.
184. ANIMAL PHYSICS. By Dr. LARDNER. Part II., Chapters VIII.—XVIII. 3s.

FINE ARTS.
20. *PERSPECTIVE FOR BEGINNERS.* Adapted to Young Students and Amateurs in Architecture, Painting, &c. By GEORGE PYNE. 2s.
40 *GLASS STAINING, AND THE ART OF PAINTING ON GLASS.* From the German of Dr. GESSERT and EMANUEL OTTO FROMBERG. With an Appendix on THE ART OF ENAMELLING. 2s. 6d.
69. *MUSIC,* A Rudimentary and Practical Treatise on. With numerous Examples. By CHARLES CHILD SPENCER. 2s. 6d.
71. *PIANOFORTE,* The Art of Playing the. With numerous Exercises & Lessons from the Best Masters. By CHARLES CHILD SPENCER. 1s. 6d.
69-71. *MUSIC & THE PIANOFORTE.* In one vol. Half bound, 5s.
181. *PAINTING POPULARLY EXPLAINED,* including Fresco, Oil, Mosaic, Water Colour, Water-Glass, Tempera, Encaustic, Miniature, Painting on Ivory, Vellum, Pottery, Enamel, Glass. &c. With Historical Sketches of the Progress of the Art by THOMAS JOHN GULLICK, assisted by JOHN TIMBS, F.S.A. Fifth Edition, revised and enlarged. 5s.‡
186. *A GRAMMAR OF COLOURING,* applied to Decorative Painting and the Arts. By GEORGE FIELD. New Edition, enlarged and adapted to the Use of the Ornamental Painter and Designer. By ELLIS A. DAVIDSON. With two new Coloured Diagrams, &c. 3s.‡
246. *A DICTIONARY OF PAINTERS, AND HANDBOOK FOR PICTURE AMATEURS;* including Methods of Painting, Cleaning, Relining and Restoring, Schools of Painting, &c. With Notes on the Copyists and Imitators of each Master. By PHILIPPE DARYL. 2s. 6d.‡

☞ *The ‡ indicates that these vols. may be had strongly bound at 6d. extra.*

7, STATIONERS' HALL COURT, LUDGATE HILL, E.C.

INDUSTRIAL AND USEFUL ARTS.

23. *BRICKS AND TILES*, Rudimentary Treatise on the Manufacture of. By E. Dobson, M.R.I.B.A. Illustrated, 3s.‡
67. *CLOCKS, WATCHES, AND BELLS*, a Rudimentary Treatise on. By Sir Edmund Beckett, LL.D., Q.C. Seventh Edition, revised and enlarged. 4s. 6d. limp; 5s. 6d. cloth boards.
83**. *CONSTRUCTION OF DOOR LOCKS*. Compiled from the Papers of A. C. Hobbs, and Edited by Charles Tomlinson, F.R.S. 2s. 6d.
162. *THE BRASS FOUNDER'S MANUAL*; Instructions for Modelling, Pattern-Making, Moulding, Turning, Filing, Burnishing, Bronzing, &c. With copious Receipts, &c. By Walter Graham. 2s.‡
205. *THE ART OF LETTER PAINTING MADE EASY*. By J. G. Badenoch. Illustrated with 12 full-page Engravings of Examples. 1s. 6d.
215. *THE GOLDSMITH'S HANDBOOK*, containing full Instructions for the Alloying and Working of Gold. By George E. Gee, 3s.‡
225. *THE SILVERSMITH'S HANDBOOK*, containing full Instructions for the Alloying and Working of Silver. By George E. Gee. 3s.;
** *The two preceding Works, in One handsome Vol., half-bound, entitled* "The Goldsmith's & Silversmith's Complete Handbook," 7s.
249. *THE HALL-MARKING OF JEWELLERY PRACTICALLY CONSIDERED*. By George E. Gee. 3s.‡
224. *COACH BUILDING*, A Practical Treatise, Historical and Descriptive. By J. W. Burgess. 2s. 6d.‡
235. *PRACTICAL ORGAN BUILDING*. By W. E. Dickson, M.A., Precentor of Ely Cathedral. Illustrated. 2s. 6d.‡
262. *THE ART OF BOOT AND SHOEMAKING*, including Measurement, Last-fitting, Cutting-out, Closing and Making. By John Bedford Leno. Numerous Illustrations. Third Edition. 2s.
263. *MECHANICAL DENTISTRY*: A Practical Treatise on the Construction of the Various Kinds of Artificial Dentures, with Formulæ, Tables, Receipts, &c. By Charles Hunter. Third Edition. 3s.‡

MISCELLANEOUS VOLUMES.

36. *A DICTIONARY OF TERMS used in ARCHITECTURE, BUILDING, ENGINEERING, MINING, METALLURGY, ARCHÆOLOGY, the FINE ARTS, &c.* By John Weale. Fifth Edition. Revised by Robert Hunt, F.R.S. Illustrated. 5s. limp; 6s. cloth boards.
50. *THE LAW OF CONTRACTS FOR WORKS AND SERVICES.* By David Gibbons. Third Edition, enlarged. 3s.‡
112. *MANUAL OF DOMESTIC MEDICINE.* By R. Gooding, B.A., M.D. A Family Guide in all Cases of Accident and Emergency. 2s.‡
112*. *MANAGEMENT OF HEALTH.* A Manual of Home and Personal Hygiene. By the Rev. James Baird, B.A. 1s.
150. *LOGIC*, Pure and Applied. By S. H. Emmens. 1s. 6d.
153. *SELECTIONS FROM LOCKE'S ESSAYS ON THE HUMAN UNDERSTANDING.* With Notes by S. H. Emmens. 2s.
154. *GENERAL HINTS TO EMIGRANTS.* 2s.
157. *THE EMIGRANT'S GUIDE TO NATAL.* By Robert James Mann, F.R.A.S., F.M.S. Second Edition. Map. 2s.
193. *HANDBOOK OF FIELD FORTIFICATION.* By Major W. W. Knollys, F.R.G.S. With 163 Woodcuts. 3s.‡
194. *THE HOUSE MANAGER*: Being a Guide to Housekeeping. Practical Cookery, Pickling and Preserving, Household Work, Dairy Management, &c. By An Old Housekeeper. 3s. 6d.‡
194, *HOUSE BOOK (The).* Comprising:—I. The House Manager. 112 By an Old Housekeeper. II. Domestic Medicine. By R. Gooding, M.D. 112*. III. Management of Health. By J. Baird. In One Vol., half-bound, 6s.

☞ *The ‡ indicates that these vols. may be had strongly bound at 6d. extra.*

LONDON : CROSBY LOCKWOOD AND SON.

EDUCATIONAL AND CLASSICAL SERIES.

HISTORY.

1. **England, Outlines of the History of;** more especially with reference to the Origin and Progress of the English Constitution. By WILLIAM DOUGLAS HAMILTON, F.S.A., of Her Majesty's Public Record Office. 4th Edition, revised. 5s.; cloth boards, 6s.
5. **Greece, Outlines of the History of;** in connection with the Rise of the Arts and Civilization in Europe. By W. DOUGLAS HAMILTON, of University College, London, and EDWARD LEVIEN, M.A., of Balliol College, Oxford. 2s. 6d.; cloth boards, 3s. 6d.
7. **Rome, Outlines of the History of:** from the Earliest Period to the Christian Era and the Commencement of the Decline of the Empire. By EDWARD LEVIEN, of Balliol College, Oxford. Map, 2s. 6d.; cl. bds. 3s. 6d.
9. **Chronology of History, Art, Literature, and Progress,** from the Creation of the World to the Present Time. The Continuation by W. D. HAMILTON, F.S.A. 3s.; cloth boards, 3s. 6d.
50. **Dates and Events in English History,** for the use of Candidates in Public and Private Examinations. By the Rev. E. RAND. 1s.

ENGLISH LANGUAGE AND MISCELLANEOUS.

11. **Grammar of the English Tongue,** Spoken and Written. With an Introduction to the Study of Comparative Philology. By HYDE CLARKE, D.C.L. Fourth Edition. 1s. 6d.
12. **Dictionary of the English Language,** as Spoken and Written. Containing above 100,000 Words. By HYDE CLARKE, D.C.L. 3s. 6d.; cloth boards, 4s. 6d.; complete with the GRAMMAR, cloth bds., 5s. 6d.
48. **Composition and Punctuation,** familiarly Explained for those who have neglected the Study of Grammar. By JUSTIN BRENAN. 18th Edition. 1s. 6d.
49. **Derivative Spelling-Book:** Giving the Origin of Every Word from the Greek, Latin, Saxon, German, Teutonic, Dutch, French, Spanish, and other Languages; with their present Acceptation and Pronunciation. By J. ROWBOTHAM, F.R.A.S. Improved Edition. 1s. 6d.
51. **The Art of Extempore Speaking:** Hints for the Pulpit, the Senate, and the Bar. By M. BAUTAIN, Vicar-General and Professor at the Sorbonne. Translated from the French. 8th Edition, carefully corrected. 2s. 6d.
53. **Places and Facts in Political and Physical Geography,** for Candidates in Examinations. By the Rev. EDGAR RAND, B.A. 1s.
54. **Analytical Chemistry, Qualitative and Quantitative,** a Course of. To which is prefixed, a Brief Treatise upon Modern Chemical Nomenclature and Notation. By WM. W. PINK and GEORGE E. WEBSTER. 2s.

THE SCHOOL MANAGERS' SERIES OF READING BOOKS,

Edited by the Rev. A. R. GRANT, Rector of Hitcham, and Honorary Canon of Ely; formerly H.M. Inspector of Schools.

INTRODUCTORY PRIMER, 3d.

	s. d.		s. d.
FIRST STANDARD	0 6	FOURTH STANDARD	1 2
SECOND ,,	0 10	FIFTH ,,	1 6
THIRD ,,	1 0	SIXTH ,,	1 6

LESSONS FROM THE BIBLE. Part I. Old Testament. 1s.
LESSONS FROM THE BIBLE. Part II. New Testament, to which is added THE GEOGRAPHY OF THE BIBLE, for very young Children. By Rev. C. THORNTON FORSTER. 1s. 2d. *⁎* Or the Two Parts in One Volume. 2s.

7, STATIONERS' HALL COURT, LUDGATE HILL, E.C.

FRENCH.

24. **French Grammar.** With Complete and Concise Rules on the Genders of French Nouns. By G. L. STRAUSS, Ph.D. 1s. 6d.
25. **French-English Dictionary.** Comprising a large number of New Terms used in Engineering, Mining, &c. By ALFRED ELWES. 1s. 6d.
26. **English-French Dictionary.** By ALFRED ELWES. 2s.
25,26. **French Dictionary** (as above). Complete, in One Vol., 3s.; cloth boards, 3s. 6d. *⁎* Or with the GRAMMAR, cloth boards, 4s. 6d.
47. **French and English Phrase Book:** containing Introductory Lessons, with Translations, several Vocabularies of Words, a Collection of suitable Phrases, and Easy Familiar Dialogues. 1s. 6d.

GERMAN.

39. **German Grammar.** Adapted for English Students, from Heyse's Theoretical and Practical Grammar, by Dr. G. L. STRAUSS. 1s. 6d.
40. **German Reader:** A Series of Extracts, carefully culled from the most approved Authors of Germany; with Notes, Philological and Explanatory. By G. L. STRAUSS, Ph.D. 1s.
41-43. **German Triglot Dictionary.** By N. E. S. A. HAMILTON. In Three Parts. Part I. German-French-English. Part II. English-German-French. Part III. French-German-English. 3s., or cloth boards, 4s.
41-43 & 39. **German Triglot Dictionary** (as above), together with German Grammar (No. 39), in One Volume, cloth boards, 5s.

ITALIAN.

27. **Italian Grammar,** arranged in Twenty Lessons, with a Course of Exercises. By ALFRED ELWES. 1s. 6d.
28. **Italian Triglot Dictionary,** wherein the Genders of all the Italian and French Nouns are carefully noted down. By ALFRED ELWES. Vol. 1. Italian-English-French. 2s. 6d.
30. **Italian Triglot Dictionary.** By A. ELWES. Vol. 2. English-French-Italian. 2s. 6d.
32. **Italian Triglot Dictionary.** By ALFRED ELWES. Vol. 3. French-Italian-English. 2s. 6d.
28,30, **Italian Triglot Dictionary** (as above). In One Vol., 7s. 6d
32. Cloth boards.

SPANISH AND PORTUGUESE.

34. **Spanish Grammar,** in a Simple and Practical Form. With a Course of Exercises. By ALFRED ELWES. 1s. 6d.
35. **Spanish-English and English-Spanish Dictionary.** Including a large number of Technical Terms used in Mining, Engineering, &c. with the proper Accents and the Gender of every Noun. By ALFRED ELWES 4s.; cloth boards, 5s. *⁎* Or with the GRAMMAR, cloth boards, 6s.
55. **Portuguese Grammar,** in a Simple and Practical Form. With a Course of Exercises. By ALFRED ELWES. 1s. 6d.
56. **Portuguese-English and English-Portuguese Dictionary.** Including a large number of Technical Terms used in Mining, Engineering, &c., with the proper Accents and the Gender of every Noun. By ALFRED ELWES. Second Edition, Revised, 5s.; cloth boards, 6s. *⁎* Or with the GRAMMAR, cloth boards, 7s.

HEBREW.

46*. **Hebrew Grammar.** By Dr. BRESSLAU. 1s. 6d.
44. **Hebrew and English Dictionary,** Biblical and Rabbinical; containing the Hebrew and Chaldee Roots of the Old Testament Post-Rabbinical Writings. By Dr. BRESSLAU. 6s.
46. **English and Hebrew Dictionary.** By Dr. BRESSLAU. 3s.
44,46. **Hebrew Dictionary** (as above), in Two Vols., complete, with
46*. the GRAMMAR, cloth boards, 12s.

LONDON: CROSBY LOCKWOOD AND SON,

LATIN.

19. **Latin Grammar.** Containing the Inflections and Elementary Principles of Translation and Construction. By the Rev. THOMAS GOODWIN, M.A., Head Master of the Greenwich Proprietary School. 1s. 6d.
20. **Latin-English Dictionary.** By the Rev. THOMAS GOODWIN, M.A. 2s.
22. **English-Latin Dictionary;** together with an Appendix of French and Italian Words which have their origin from the Latin. By the Rev. THOMAS GOODWIN, M.A. 1s. 6d.
20,22. **Latin Dictionary** (as above). Complete in One Vol., 3s. 6d. cloth boards, 4s. 6d. *** Or with the GRAMMAR, cloth boards, 5s. 6d.

LATIN CLASSICS. With Explanatory Notes in English.
1. **Latin Delectus.** Containing Extracts from Classical Authors, with Genealogical Vocabularies and Explanatory Notes, by H. YOUNG. 1s. 6d.
2. **Cæsaris** Commentarii de Bello Gallico. Notes, and a Geographical Register for the Use of Schools, by H. YOUNG. 2s.
3. **Cornelius Nepos.** With Notes. By H. YOUNG. 1s.
4. **Virgilii** Maronis Bucolica et Georgica. With Notes on the Bucolics by W. RUSHTON, M.A., and on the Georgics by H. YOUNG. 1s. 6d.
5. **Virgilii** Maronis Æneis. With Notes, Critical and Explanatory, by H. YOUNG. New Edition, revised and improved With copious Additional Notes by Rev. T. H. L. LEARY, D.C.L., formerly Scholar of Brasenose College, Oxford. 3s.
5* ——— Part 1. Books i.—vi., 1s. 6d.
5** ——— Part 2. Books vii.—xii., 2s.
6. **Horace**; Odes, Epode, and Carmen Sæculare. Notes by H. YOUNG. 1s. 6d.
7. **Horace**; Satires, Epistles, and Ars Poetica. Notes by W. BROWNRIGG SMITH, M.A., F.R.G.S. 1s. 6d.
8. **Sallustii** Crispi Catalina et Bellum Jugurthinum. Notes, Critical and Explanatory, by W. M. DONNE, B.A., Trin. Coll., Cam. 1s. 6d.
9. **Terentii** Andria et Heautontimorumenos. With Notes, Critical and Explanatory, by the Rev. JAMES DAVIES, M.A. 1s. 6d.
10. **Terentii** Adelphi, Hecyra, Phormio. Edited, with Notes, Critical and Explanatory, by the Rev. JAMES DAVIES, M.A. 2s.
11. **Terentii** Eunuchus, Comœdia. Notes, by Rev. J. DAVIES, M.A. 1s. 6d.
12. **Ciceronis** Oratio pro Sexto Roscio Amerino. Edited, with an Introduction, Analysis, and Notes, Explanatory and Critical, by the Rev. JAMES DAVIES, M.A. 1s. 6d.
13. **Ciceronis** Orationes in Catilinam, Verrem, et pro Archia. With Introduction, Analysis, and Notes, Explanatory and Critical, by Rev. T. H. L. LEARY, D.C.L. formerly Scholar of Brasenose College, Oxford. 1s. 6d.
14. **Ciceronis** Cato Major, Lælius, Brutus, sive de Senectute, de Amicitia, de Claris Oratoribus Dialogi. With Notes by W. BROWNRIGG SMITH M.A., F.R.G.S. 2s.
16. **Livy**: History of Rome. Notes by H. YOUNG and W. B. SMITH, M.A. Part 1. Books i., ii., 1s. 6d.
16*. ——— Part 2. Books iii., iv., v., 1s. 6d.
17. ——— Part 3. Books xxi., xxii., 1s. 6d.
19. **Latin Verse Selections**, from Catullus, Tibullus, Propertius, and Ovid. Notes by W. B. DONNE, M.A., Trinity College, Cambridge. 2s.
20. **Latin Prose Selections**, from Varro, Columella, Vitruvius, Seneca, Quintilian, Florus, Velleius Paterculus, Valerius Maximus Suetonius, Apuleius, &c. Notes by W. B. DONNE, M.A. 2s.
21. **Juvenalis** Satiræ. With Prolegomena and Notes by T. H. S. ESCOTT, B.A., Lecturer on Logic at King's College, London. 2s.

GREEK.

14. **Greek Grammar**, in accordance with the Principles and Philological Researches of the most eminent Scholars of our own day. By HANS CLAUDE HAMILTON. 1s. 6d.
15,17. **Greek Lexicon.** Containing all the Words in General Use, with their Significations, Inflections, and Doubtful Quantities. By HENRY R. HAMILTON. Vol. 1. Greek-English, 2s. 6d.; Vol. 2. English-Greek, 2s. Or the Two Vols. in One, 4s. 6d.; cloth boards, 5s.
14,15. **Greek Lexicon** (as above). Complete, with the GRAMMAR, in
17. One Vol., cloth boards, 6s.

GREEK CLASSICS. With Explanatory Notes in English.

1. **Greek Delectus.** Containing Extracts from Classical Authors, with Genealogical Vocabularies and Explanatory Notes, by H. YOUNG. New Edition, with an improved and enlarged Supplementary Vocabulary, by JOHN HUTCHISON, M.A., of the High School, Glasgow. 1s. 6d.
2, 3. **Xenophon's Anabasis;** or, The Retreat of the Ten Thousand. Notes and a Geographical Register, by H. YOUNG. Part 1. Books i. to iii., 1s. Part 2. Books iv. to vii., 1s.
4. **Lucian's Select Dialogues.** The Text carefully revised, with Grammatical and Explanatory Notes, by H. YOUNG. 1s. 6d.
5-12. **Homer, The Works of.** According to the Text of BAEUMLEIN. With Notes, Critical and Explanatory, drawn from the best and latest Authorities, with Preliminary Observations and Appendices, by T. H. L. LEARY, M.A., D.C.L.

THE ILIAD:	Part 1. Books i. to vi., 1s. 6d.	Part 3. Books xiii. to xviii., 1s. 6d.
	Part 2. Books vii. to xii., 1s. 6d.	Part 4. Books xix. to xxiv., 1s. 6d.
THE ODYSSEY:	Part 1. Books i. to vi., 1s. 6d.	Part 3. Books xiii. to xviii., 1s. 6d.
	Part 2. Books vii. to xii., 1s. 6d.	Part 4. Books xix. to xxiv., and Hymns, 2s.

13. **Plato's Dialogues:** The Apology of Socrates, the Crito, and the Phædo. From the Text of C. F. HERMANN. Edited with Notes, Critical and Explanatory, by the Rev. JAMES DAVIES, M.A. 2s.
14-17. **Herodotus, The History of**, chiefly after the Text of GAISFORD. With Preliminary Observations and Appendices, and Notes, Critical and Explanatory, by T. H. L. LEARY, M.A., D.C.L.
　　Part 1. Books i., ii. (The Clio and Euterpe), 2s.
　　Part 2. Books iii., iv. (The Thalia and Melpomene), 2s.
　　Part 3. Books v.-vii. (The Terpsichore, Erato, and Polymnia), 2s.
　　Part 4. Books viii., ix. (The Urania and Calliope) and Index, 1s. 6d.
18. **Sophocles:** Œdipus Tyrannus. Notes by H. YOUNG. 1s.
20. **Sophocles:** Antigone. From the Text of DINDORF. Notes, Critical and Explanatory, by the Rev. JOHN MILNER, B.A. 2s.
23. **Euripides:** Hecuba and Medea. Chiefly from the Text of DINDORF. With Notes, Critical and Explanatory, by W. BROWNRIGG SMITH, M.A., F.R.G.S. 1s. 6d.
26. **Euripides:** Alcestis. Chiefly from the Text of DINDORF. With Notes, Critical and Explanatory, by JOHN MILNER, B.A. 1s. 6d.
30. **Æschylus:** Prometheus Vinctus: The Prometheus Bound. From the Text of DINDORF. Edited, with English Notes, Critical and Explanatory, by the Rev. JAMES DAVIES, M.A. 1s.
32. **Æschylus:** Septem Contra Thebes: The Seven against Thebes. From the Text of DINDORF. Edited, with English Notes, Critical and Explanatory, by the Rev. JAMES DAVIES, M.A. 1s.
40. **Aristophanes:** Acharnians. Chiefly from the Text of C. H. WEISE. With Notes, by C. S. T. TOWNSHEND, M.A. 1s. 6d.
41. **Thucydides:** History of the Peloponnesian War. Notes by H. YOUNG. Book 1. 1s. 6d.
42. **Xenophon's Panegyric on Agesilaus.** Notes and Introduction by LL. F. W. JEWITT. 1s. 6d.
43. **Demosthenes.** The Oration on the Crown and the Philippics. With English Notes. By Rev. T. H. L. LEARY, D.C.L., formerly Scholar of Brasenose College, Oxford. 1s. 6d.

CROSBY LOCKWOOD AND SON, 7, STATIONERS' HALL COURT, E.C.

www.ingramcontent.com/pod-product-compliance
Lightning Source LLC
Chambersburg PA
CBHW032042230426
43672CB00009B/1444